The Biology
Student's Self-test
Coloring Book

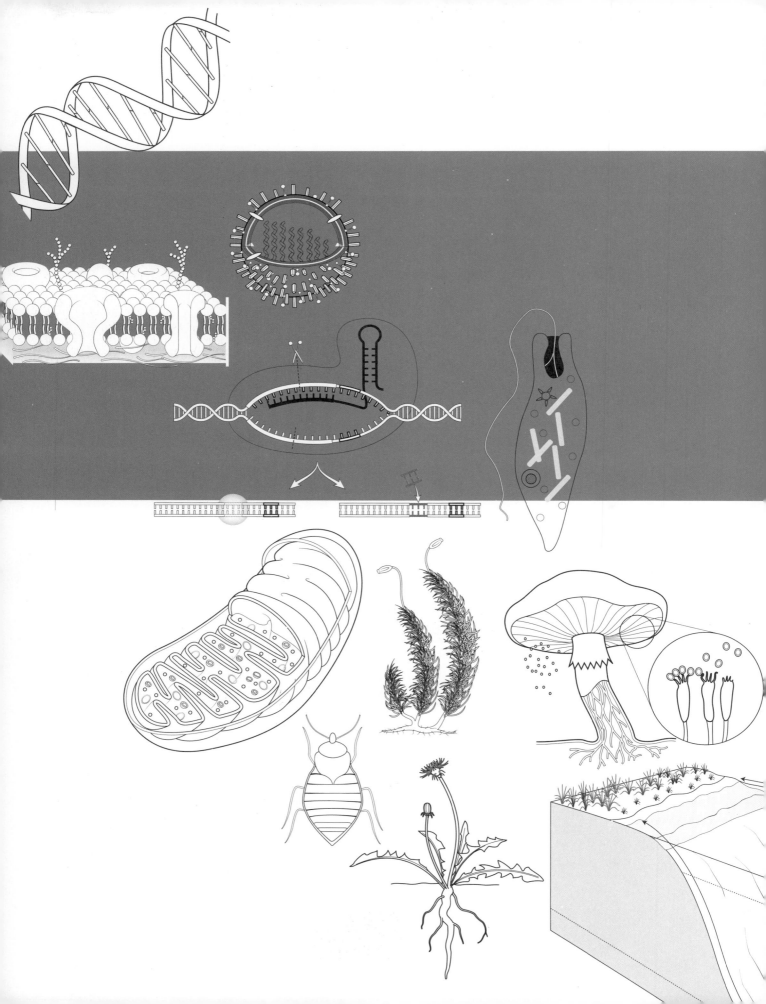

The Biology
Student's Self-test
Coloring Book

René Fester Kratz, Ph.D.

Everett Community College,
Everett, Washington

BARRON'S

First edition for North America published in 2019
by Kaplan, Inc.

Conceived, designed, and produced by The Bright Press, an imprint of the Quarto Group
58 West Street, Brighton, BN1 2RA, United Kingdom

Publisher: Mark Searle
Creative Director: James Evans
Development Editor: Jacqui Sayers
Project Editor and Designer: D & N Publishing, Baydon, Wiltshire, UK
Illustrator: Medical Artist Ltd (*www.medical-artist.com*)

Published by Barron's Educational Series, Inc.
750 Third Avenue
New York, NY 10017
www.barronseduc.com

ISBN: 978-1-4380-1231-5

Printed in China

9 8 7 6 5 4 3 2 1

Contents

Introduction

From bacteria to blue herons (*Ardea herodias*), and from kookaburras to Komodo dragons (*Varanus komodoensis*), living creatures blanket every nook and cranny of the earth's surface. This layer of life, called the biosphere, extends 40 miles (64 km) up into the atmosphere and 12 miles (19 km) deep into the crust. Humans are members of this complex web of life, interacting directly and indirectly with other species to share the resources we all need to survive. Understanding what each species needs, how it grows and develops, and how it interacts with other species is all part of the science of biology. Learning about biology can help us understand our own species and our place in the world.

A solid foundation in the biological sciences is essential for careers in health care and environmental management. Studying biology can also lead to a better appreciation for nature and an improved understanding of complex issues like biodiversity and climate change. Whatever your reason for studying biology, this book provides an active learning experience that will help you master the subject, from the workings of microscopic cells to ecological interactions on a global scale. On each page, you can quiz yourself by filling in the names of structures and color-coding major elements that are similar from one illustration to the next. Labeling figures and diagrams yourself will help you remember details much better than if you simply review them in a textbook. The unique connection between hand, eye, and mind makes this biology coloring book a fun and helpful study tool. It is accessible to all those looking for a different approach to studying and to expand their knowledge of the subject.

How This Book Is Organized

Featuring more than 200 computer-rendered line drawings in a clean design, this book is divided into 15 comprehensive chapters. It covers the fundamental concepts of biology—from cells and DNA science to ecology and evolution. It introduces the major groups of organisms and the important principles of plant and animal physiology. It also has an in-depth review of human anatomy and physiology. As you work your way through the book, you will gain a broad understanding of the most important concepts of biology and an appreciation for the diversity of life on Earth.

How to Use This Book

This book is designed to help students and professionals identify important structures and processes in biology. The colored leader lines aid the process by clearly pointing out each feature. The functions of coloring and labeling allow you to practice your knowledge of the major elements of each process and the structural details of organisms. Coloring is best done using either pencils or ballpoint pens (not felt-tip pens) in a variety of dark and light colors. Where possible, you should use the same color for like structures, so that all completed illustrations can be utilized later as visual references. For the section on human anatomy and physiology, you can apply the conventional coloring scheme of green for lymphatic structures, yellow for nerves, red for arteries, and blue for veins. Labeling the colored leader lines that point to separate parts of the illustration enables you to test and then check your knowledge using the answers that are printed at the bottom of the page.

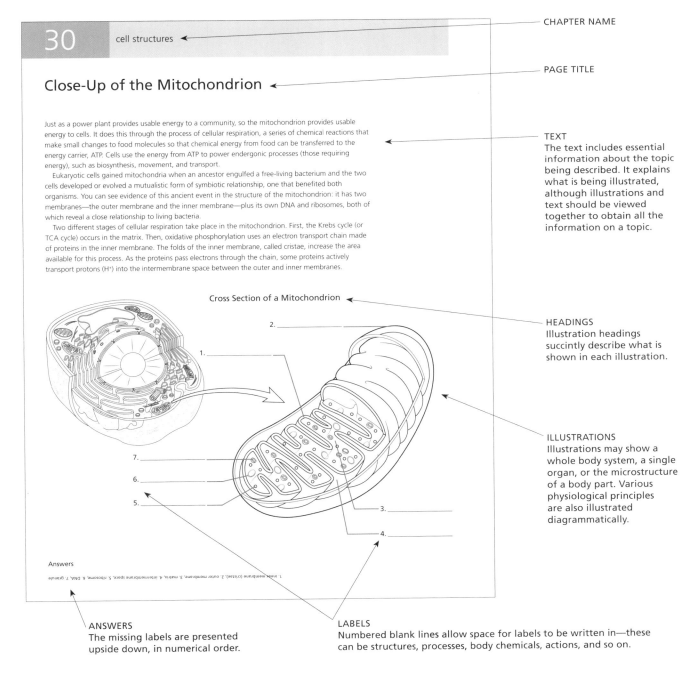

30   cell structures ← **CHAPTER NAME**

PAGE TITLE

Close-Up of the Mitochondrion ←

Just as a power plant provides usable energy to a community, so the mitochondrion provides usable energy to cells. It does this through the process of cellular respiration, a series of chemical reactions that make small changes to food molecules so that chemical energy from food can be transferred to the energy carrier, ATP. Cells use the energy from ATP to power endergonic processes (those requiring energy), such as biosynthesis, movement, and transport.

    Eukaryotic cells gained mitochondria when an ancestor engulfed a free-living bacterium and the two cells developed or evolved a mutualistic form of symbiotic relationship, one that benefited both organisms. You can see evidence of this ancient event in the structure of the mitochondrion: it has two membranes—the outer membrane and the inner membrane—plus its own DNA and ribosomes, both of which reveal a close relationship to living bacteria.

    Two different stages of cellular respiration take place in the mitochondrion. First, the Krebs cycle (or TCA cycle) occurs in the matrix. Then, oxidative phosphorylation uses an electron transport chain made of proteins in the inner membrane. The folds of the inner membrane, called cristae, increase the area available for this process. As the proteins pass electrons through the chain, some proteins actively transport protons (H^+) into the intermembrane space between the outer and inner membranes.

TEXT
The text includes essential information about the topic being described. It explains what is being illustrated, although illustrations and text should be viewed together to obtain all the information on a topic.

Cross Section of a Mitochondrion ←

HEADINGS
Illustration headings succintly describe what is shown in each illustration.

ILLUSTRATIONS
Illustrations may show a whole body system, a single organ, or the microstructure of a body part. Various physiological principles are also illustrated diagrammatically.

Answers

1. inner membrane (cristae). 2. outer membrane. 3. matrix. 4. intermembrane space. 5. ribosome. 6. DNA. 7. granule

ANSWERS
The missing labels are presented upside down, in numerical order.

LABELS
Numbered blank lines allow space for labels to be written in—these can be structures, processes, body chemicals, actions, and so on.

Characteristics of Living Things

The same chemical elements make up the structures of all living (biotic) and nonliving (abiotic) things on planet Earth, yet living things have many unique properties. Biologists have spent a great deal of time discussing these characteristics and trying to pinpoint exactly which ones most precisely define life. You can find many versions of these lists, but they all have certain common traits at their core. Some of these traits may be found individually in abiotic materials, but only living things show them all.

So, what makes living things unique? Biologists agree that all living things are made of cells and that they grow and can reproduce, either sexually or asexually. By giving copies of their genetic material in the form of chromosomes to their offspring, parents pass on their inheritable traits. Occasional changes can occur through mutation, allowing individuals and, hence, populations of organisms to adapt to changes in their environment over long periods of evolutionary time. Staying alive requires living things to respond to stimuli in the environment and to maintain an internal balance, called homeostasis. All organisms require energy in order to grow, move, reproduce, and maintain their organization. Living things also exchange matter with their environment, taking in materials like food and releasing wastes.

Characteristics That Define Living Things

1. _____

2. _____

3. _____

4. _____

5. _____

6. _____

7. _____

8. _____

Answers

1. made of cells, 2. respond to stimuli, 3. grow, 4. require energy, 5. maintain homeostasis, 6. display heredity, 7. exchange matter, 8. reproduce

Organization of Living Things

Biologists study life at many different levels of organization. Biologists call each living thing an organism; some organisms, like bacteria, are no bigger than a single cell, while others, like a plant or an animal, may have billions or trillions of cells. Cell and molecular biologists focus on the smallest unit of life, the cell, and the nonliving atoms and molecules that form the structures in cells. Biologists interested in anatomy and physiology study how different cell types combine to form tissues and how layers of tissues form organs. Organs form organ systems, which together regulate the physiology of multicellular organisms.

Environmental scientists and ecologists look at how organisms interact with each other and with their environment. A group of organisms of one type living in the same place is a population. When several populations live in the same area, they form a community. Communities interacting with their physical environment form ecosystems, the largest of which is the biosphere. The biosphere rises up into the atmosphere and down into the soil, forming the layer of life that spans the entire surface of Earth, making us unique among all the planets discovered to date.

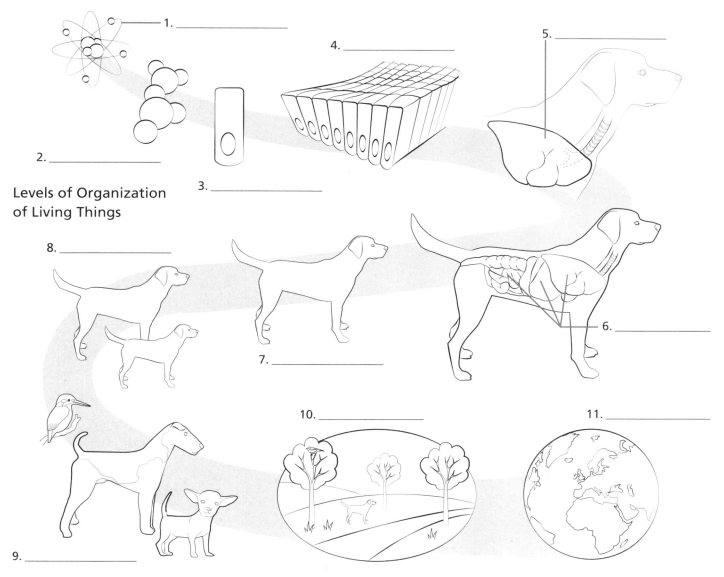

1. _____

4. _____

5. _____

2. _____

3. _____

Levels of Organization of Living Things

8. _____

7. _____

6. _____

9. _____

10. _____

11. _____

Answers

1. atom, 2. molecule, 3. cell, 4. tissue, 5. organ, 6. organ system, 7. organism, 8. population, 9. community, 10. ecosystem, 11. biosphere

Flow of Genetic Information

Deoxyribonucleic acid (DNA) contains the instructions required by living organisms to build and control all the molecules necessary for the structure and function of cells. The information is encoded in a chemical pattern of deoxyribonucleotides that contains four different nitrogenous bases: adenine (A), guanine (G), thymine (T), and cytosine (C). These nucleotides join together to form long chains; it's the order of the bases in the chains that spells out the chemical code in DNA. To form a complete DNA molecule, two partner strands with complementary codes attach to each other, twisting into the familiar double helix of DNA. When cells reproduce, they use DNA replication to make copies of their chromosomes so that each resulting cell has a complete set of instructions. DNA replication is semiconservative, meaning that cells use each strand of the double helix as a template to make a new partner strand, producing DNA molecules that are half original and half new.

When cells need molecules, such as enzymes or ribonucleic acid (RNA), to perform a particular function, they use transcription to copy the relevant code from their DNA into a complementary molecule of messenger RNA (mRNA). Like DNA, RNA also encodes information in a chemical pattern of four alternating ribonucleotides, but it contains the nitrogenous base uracil (U) instead of thymine (T). Transcription can produce several different kinds of RNA, including mRNA, transfer RNA (tRNA), and ribosomal RNA (rRNA).

Cells use the code in mRNA as the pattern for the construction of the polypeptide chains necessary for all proteins, including enzymes. The code is translated in units of three nucleotides called codons, each of which represents one amino acid. tRNA molecules supply the amino acids, using their anticodons to pair with codons in the mRNA so that each amino acid is placed correctly in the polypeptide chain. Translation occurs on ribosomes, which are made of rRNA and protein.

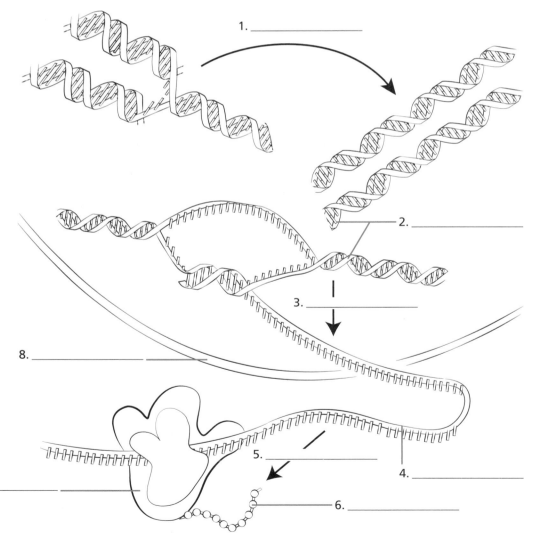

1. _____

2. _____

3. _____

4. _____

5. _____

6. _____

7. _____

8. _____

Flow of Genetic Information

The Tree of Life

Biologists compare the structures, chemistry, and genetic material of living things in order to decide how organisms relate to each other through evolutionary time. These relationships can be shown by the construction of phylogenetic trees, in which nodes indicate common ancestors and branch lengths demonstrate time since evolutionary divergence. The modern system of classification is based on the work of Carl Linnaeus (1707–1778), a Swedish naturalist who placed organisms into hierarchical categories of relationship and who developed the binomial system of naming species that's still in use today.

A major revision to the modern understanding of phylogeny occurred in the 1970s, when microbiologist Carl Woese (1928–2012) compared the sequences of genes coding for rRNA among many different organisms. Woese's work revealed that prokaryotes should be split into two groups, rather than kept in one as had been assumed based on cell structure. Because of this, biologists now recognize three major branches on the tree of life: the domains Archaea, Bacteria, and Eukarya. Within these domains, biologists recognize the subcategories of kingdom, phylum, class, order, family, genus, and species. A proper scientific name for a particular species consists of the genus and specific epithet (species name) and must be italicized or underlined. For example, humans are *Homo sapiens*, and the tiger is *Panthera tigris*.

Evolutionary Relationships of Life and the System of Classification

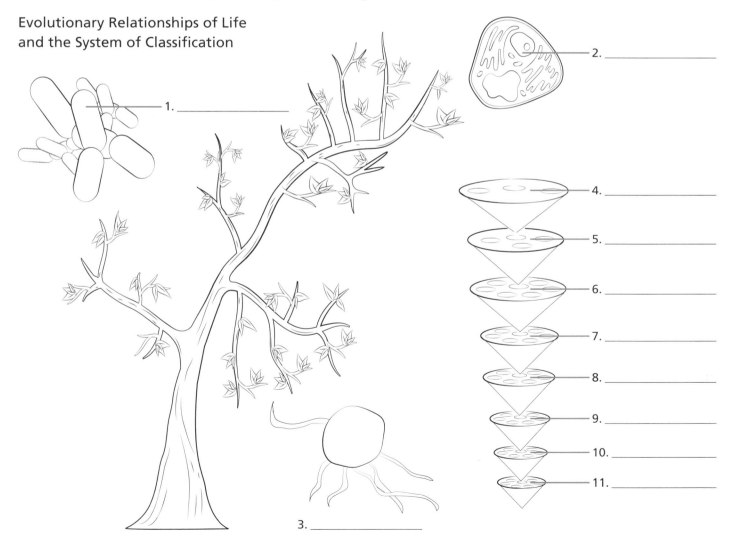

1. _____

2. _____

3. _____

4. _____

5. _____

6. _____

7. _____

8. _____

9. _____

10. _____

11. _____

Answers

1. Bacteria, 2. Eukarya, 3. Archaea, 4. domain, 5. kingdom, 6. phylum, 7. class, 8. order, 9. family, 10. genus, 11. species

The Process of Science

Science is a process for gathering information about the natural world. Scientists observe the world using their five senses but may extend these with the use of equipment like telescopes, microscopes, or probes. Scientists ask questions about what they perceive and seek to understand the laws and processes that govern the natural world through repeated cycles of testing and observation. They propose explanations called hypotheses for how things might work, and then make predictions and perform tests to challenge these mental models. They collect the results or data from their experiments and analyze it, making conclusions about whether their hypotheses were correct or whether they need to be revised.

Scientists share their work with other scientists all around the world, comparing results and collaborating on new experiments. Scientists submit their work to scientific journals for publication. In order for their work to be published in the highest-quality scientific journals, it must pass peer review, which is a critical review process conducted by other scientists in the same field. Over time, the work of many scientists combines to form scientific theories. Unlike hypotheses, which are tentative explanations of how the world works, scientific theories are typically supported by multiple lines of evidence from several sources. Although theories can change as scientists gather new information, scientists typically regard them as highly probable explanations.

Process of Science

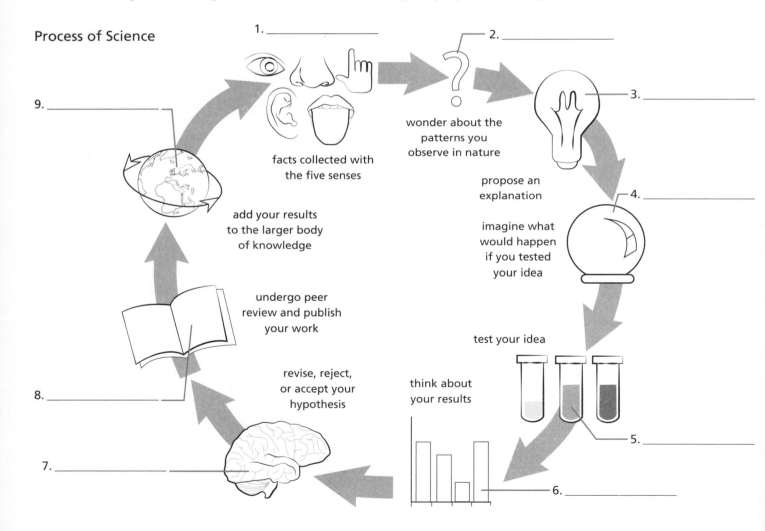

1. _____

2. _____

3. _____

4. _____

5. _____

6. _____

7. _____

8. _____

9. _____

wonder about the patterns you observe in nature

facts collected with the five senses

propose an explanation

imagine what would happen if you tested your idea

add your results to the larger body of knowledge

test your idea

undergo peer review and publish your work

think about your results

revise, reject, or accept your hypothesis

Answers

1. observation, 2. question, 3. hypothesis, 4. prediction, 5. experiment, 6. analyze, 7. conclude, 8. communicate, 9. contribute

Performing a Controlled Experiment

Controlled experiments are an important tool scientists use to test their ideas. Scientists design these experiments very carefully so that all of the conditions the subjects will experience are standardized, with the exception of the condition that the scientist wants to test. Scientists call these conditions variables because they have the potential to be different. Any condition that the scientist keeps the same between groups of subjects is called a controlled variable (it isn't allowed to differ), and the condition that the scientist changes on purpose is called the experimental variable. The scientist observes the changes that occur during the experiment and collects measurements, or data, on these changes. The changes that the scientist measures are called the responding variables. The experimental variable is also sometimes called the independent or manipulated variable, and the responding variable is also called the dependent variable.

In addition to using precise terminology for variables, scientists also have names for the groups of test subjects in an experiment. The experimental group is the group of subjects that receives the experimental variable, or that is exposed to different variations of the experimental variable. A control group is a group of subjects that is kept the same as the experimental group in every way except for the difference in the experimental variable. Control groups are often exposed to standard or natural conditions as a comparison for the effect of the changed condition of the experimental variable.

Example of a Controlled Experiment

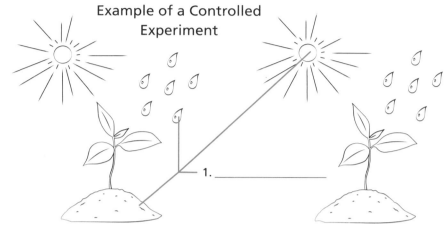

1. _____

plant two groups of plants in identical conditions

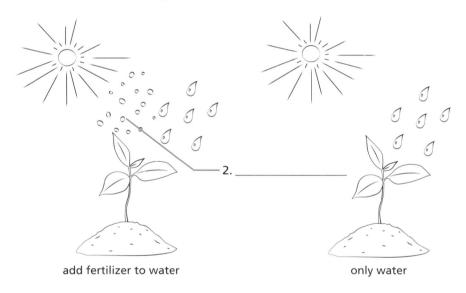

2. _____

add fertilizer to water

only water

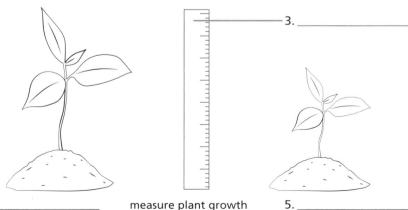

3. _____

4. _____

measure plant growth

5. _____

Answers

Atoms and Molecules

Bohr Models

Ball and Stick Models

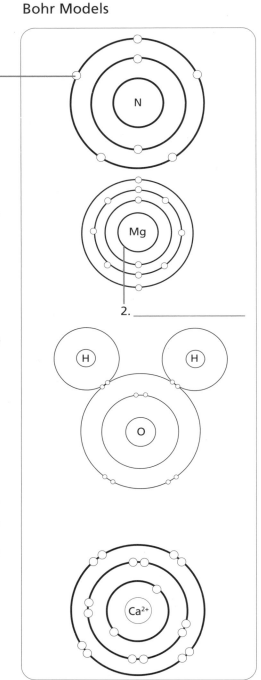

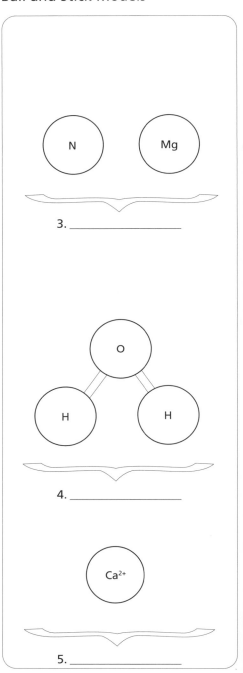

1. _____

2. _____

3. _____

4. _____

5. _____

Atoms are the smallest units of matter that retain the properties of an element. Each atom is composed of smaller particles. Protons and neutrons make up the center, or nucleus, of the atom. Each of these particles has a mass of 1. Protons have a positive charge, while neutrons have no charge. Electrons are negatively charged particles that orbit the nucleus of the atom. Electrons have so little mass that they do not contribute significantly to the mass of the atom. Each atom has the same number of electrons as it does protons, so that the positive and negative charges balance each other and the atom has no net charge.

Electrons orbit the nucleus in regions called electron shells, each of which can hold a certain number of electrons. If an atom's outermost electron shell isn't full, it will react with other atoms in ways that let it fill the outer shell. One type of interaction is when atoms join together with other atoms to form molecules, compounds made of two or more atoms. Another way an atom can fill its outermost shell is to take electrons from other atoms. An atom that gains electrons will become a negatively charged ion, while an atom that loses electrons becomes positively charged.

Answers

Chemical Bonds

Chemical bonds are the attractions that hold atoms or chemical groups together. Ions with opposite charges are attracted to each other by ionic bonds. In a dry chemical, like table salt, ionic bonds are very strong, but in the watery environment of cells, they are weak.

When atoms share electrons with each other, they form covalent bonds. Covalent bonds are the strong bonds that hold together the carbon skeletons of molecules that make up cells. Each shared pair of electrons is one covalent bond. Some atoms will share more than one pair of electrons, forming double or even triple covalent bonds. When atoms share electrons very equally with each other, the electrical charge around the bond is neutral and the chemical group is nonpolar. Some atoms have more pull for electrons than others do, however, resulting in unequal sharing and a polar covalent bond. Polar covalent bonds have a slightly uneven distribution of charge around the bond.

Polar covalent bonds set up conditions that allow hydrogen bonds, which are weak electrical attractions between groups that have slight positive charges and those that have slight negative charges. Many of the unique properties of water, such as surface tension and cohesion, result from the hydrogen bonds between individual water molecules.

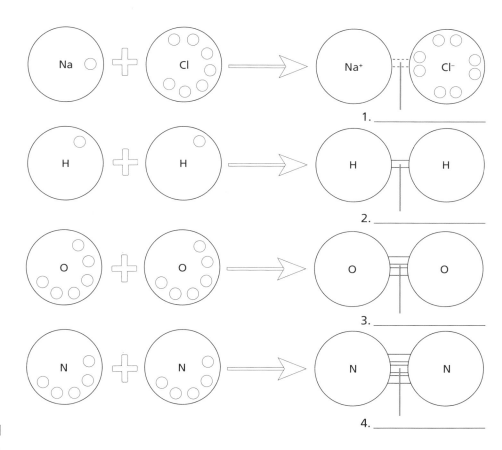

Chemical Bond Types That Hold Atoms and Molecules Together

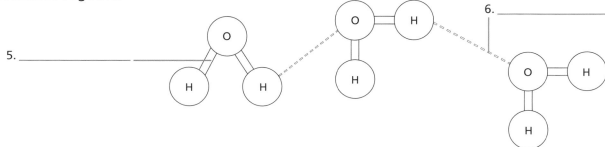

Functional Groups

Large molecules, called macromolecules, make up the structure of cells. They have a skeleton that is primarily composed of carbon atoms. What makes the properties of each type of macromolecule different from others is the small chemical groups that are attached to this skeleton. These functional groups can help you both predict the properties of a molecule and also identify it.

Proteins contain many amino and carboxyl groups. Amino groups consist of one nitrogen atom and two hydrogen atoms. Carboxyl groups contain a carbon atom, two oxygen atoms, and a hydrogen atom. The carbon is double-bonded to one of the oxygen atoms and single-bonded to a hydroxyl group. Hydroxyl groups contain an oxygen and a hydrogen atom. Proteins can also contain sulfhydryl groups made of a sulfur atom that is bonded to a hydrogen atom.

Sugars contain carbonyl and hydroxyl groups. Carbonyl groups consist of a carbon atom that is double-bonded to an oxygen atom. If the carbon is also bonded to at least one hydrogen atom, then the molecule is an aldehyde. If not, the molecule is a ketone.

Nucleic acids and phospholipids contain phosphate groups, which have four oxygen atoms bound to a central phosphorous atom. They carry a negative electrical charge.

Functional Group	Structural Formula	Ball and Stick Model
1. _____		
2. _____		
3. _____		
4. _____		
5. _____		
6. _____		
7. _____		

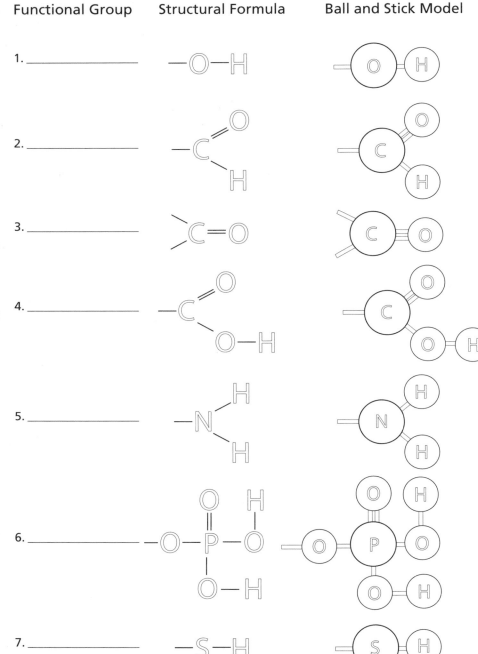

Answers

Monomers and Polymers

Three of the four groups of macromolecules that form the primary structures of cells contain polymers—long-chain molecules made of repeating subunits called monomers. The monomer for each group is different, but the mechanism by which monomers are joined together and then split apart is the same.

Cells join monomers together by condensation reactions (also called dehydration synthesis). During a condensation reaction, a monomer joins a growing polymer and a water molecule leaves. The water molecule forms when a hydroxyl group from the incoming monomer combines with a hydrogen atom from the polymer. These join together at the same time that a covalent bond forms between an atom in the monomer and one in the polymer.

Cells also split polymers apart to release monomers from a polymer. This process, called hydrolysis, is essentially the reverse of condensation: a water molecule enters the reaction, and the monomer is freed from the chain. During hydrolysis, the water molecule splits as the bond between the monomer and the rest of the chain breaks. A hydrogen atom attaches to the polymer, and the remaining hydroxyl group attaches to the monomer as it goes free.

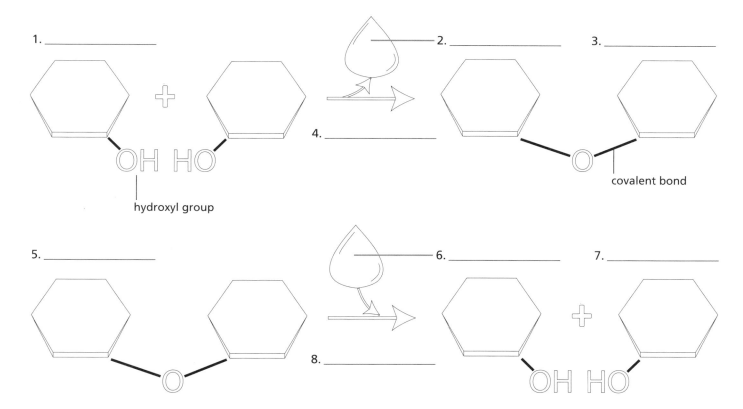

1. _____

2. _____

3. _____

4. _____

hydroxyl group

covalent bond

5. _____

6. _____

7. _____

8. _____

Condensation and Hydrolysis

Answers

Carbohydrates

Carbohydrates are an important source of energy for cells and can also play a role in structural support and signaling. Carbon, hydrogen, and oxygen atoms combine to form carbohydrate molecules. In their smallest form, carbohydrates are monosaccharides, or simple sugars. Three to seven carbon atoms form the carbon skeleton of the molecule, and, in addition to a hydrogen atom, a hydroxyl group is attached to every carbon atom but one, which has a carbonyl group attached to it. In water, monosaccharides adopt a ring structure most of the time.

Two monosaccharides may bond together through condensation to form disaccharides, or even more can combine to form long chains called polysaccharides, or complex carbohydrates. Glucose molecules bind in different ways to form three polysaccharides that are very important to cells: starch, cellulose, and glycogen. Plants make starch to store energy and matter, and many organisms—including humans—depend upon this starch as a source of food. In addition to chains, starch can also form in a spiral structure. Plants make cellulose to provide structure to their cell walls. Because of the way the glucose molecules join together to form cellulose, humans can't digest it, although it still plays an important role as fiber in our digestive systems. Animals join glucose to form glycogen, then store it in muscle cells and the liver as short-term energy reserves.

Types of Carbohydrates

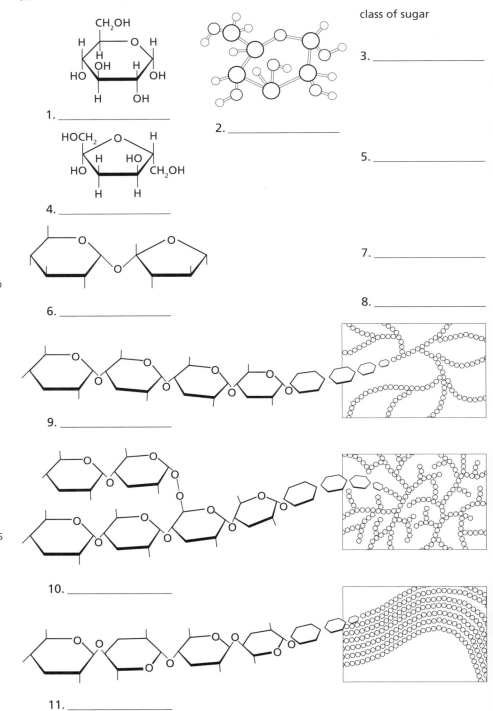

class of sugar

1. _____

2. _____

3. _____

4. _____

5. _____

6. _____

7. _____

8. _____

9. _____

10. _____

11. _____

Answers

Lipids

Lipids are a diverse group of molecules that have one thing in common: they are hydrophobic, meaning they don't mix well with water. Many organisms store energy and matter reserves as lipids, such as fats and oils. In animals, fats also provide insulation and cushioning for organs. Some organisms produce lipids like waxes as a way to waterproof certain structures.

Carbon and hydrogen atoms bond together in a variety of arrangements to make the carbon skeleton of lipids. Fats and oils are triglycerides, which form from the condensation of one glycerol molecule and three fatty acids. Saturated fatty acids, like those in animal fats, contain only single covalent bonds between their carbon atoms. Unsaturated fatty acids, like those in plant and fish oils, contain some double covalent bonds. Because they are straight, saturated fatty acids pack more tightly, making saturated fats solid at room temperature, whereas the crooked structure of unsaturated oils keeps them looser and liquid.

Phospholipids are a main component in the plasma membranes of all cells. Structurally similar to triglycerides, they form from condensation of a glycerol molecule, two fatty acid chains, and a phosphate-containing head group. The head group contains positive and negative charges, making the head group hydrophilic, or attracted to water. Overall, phospholipids have a dual nature due to their hydrophilic heads and hydrophobic tails.

Sterols are lipids composed of four fused rings. An example is cholesterol, an important component of the plasma membranes of animal cells. Some vitamins and important hormones are also sterols.

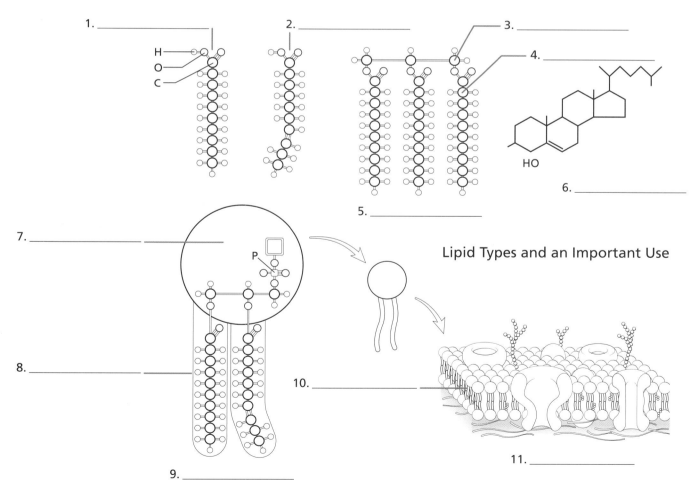

Lipid Types and an Important Use

Answers

Proteins

Proteins provide many important functions to cells. They act as enzymes, speeding up chemical reactions. Cytoskeletal proteins provide structure to cells and facilitate movement of materials within them. Movement of cytoskeletal proteins allows muscle cells to contract and sperm to swim. Defensive proteins called antibodies protect organisms from infection. Proteins can also be involved in signaling between cells, either by acting as receptors for signals or, like the protein hormone insulin, as the signals themselves. In addition, proteins also regulate gene expression and thus play crucial roles in both the development and maintenance of homeostasis.

The monomer of proteins is the amino acid. Amino acids combine by condensation to form peptide chains held together by covalent bonds called peptide bonds. Twenty different amino acids exist in cells, each having the same core structure of a central carbon atom, with an amino group and a carboxyl group attached on either side. The central carbon atom also attaches to a variable group, called the R group or side chain, which distinguishes the 20 amino acids from one another. The properties of the 20 amino acids vary depending upon the functional groups found in their R group: R groups can be hydrophilic, hydrophobic, polar, nonpolar, acidic, or basic. Amino acids join together to form chains called polypeptides. The properties of the amino acids in the polypeptide chain affect how it folds and how it functions.

Amino Acid Types

1. _____ 2. _____

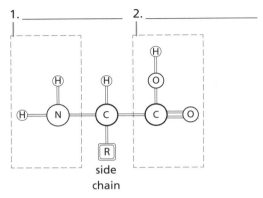

side chain

3. _____

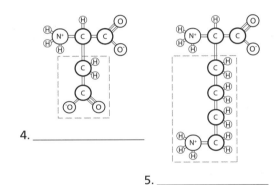

4. _____   5. _____

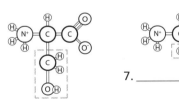

7. _____

6. _____

Polypeptide

8. _____ terminus _____

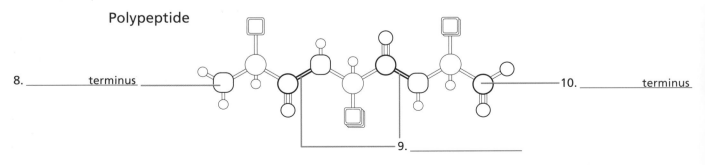

10. _____ terminus

9. _____

Answers

Close-Up of Protein Structure

One or more polypeptide chains fold together to form a functional protein. Each protein has a particular shape that is essential to its function. Ultimately, it's the number and composition of amino acids in a polypeptide chain that determine how it folds. Interactions between the chemical groups of the amino acids trigger different types of folds and can also bond multiple polypeptide chains together. If proteins denature (unfold), they cease to function and cells may die.

Biologists separate elements of protein structure into four different categories. The primary structure of a protein is the amino acid sequence of the polypeptide chain, which is held together by peptide bonds. Hydrogen bonds between R groups can trigger localized folds called α-helices or β-strands, which represent the secondary structure of a protein. Other interactions between R groups, including ionic bonds, covalent bonds, and nonpolar interactions, cause the entire polypeptide chain to form into its unique shape, or conformation. This three-dimensional shape is the tertiary structure of the protein. Finally, if more than one polypeptide chain joins to others to form the functional protein, the protein has quaternary structure. Quaternary structure is also held together by interactions between the R groups.

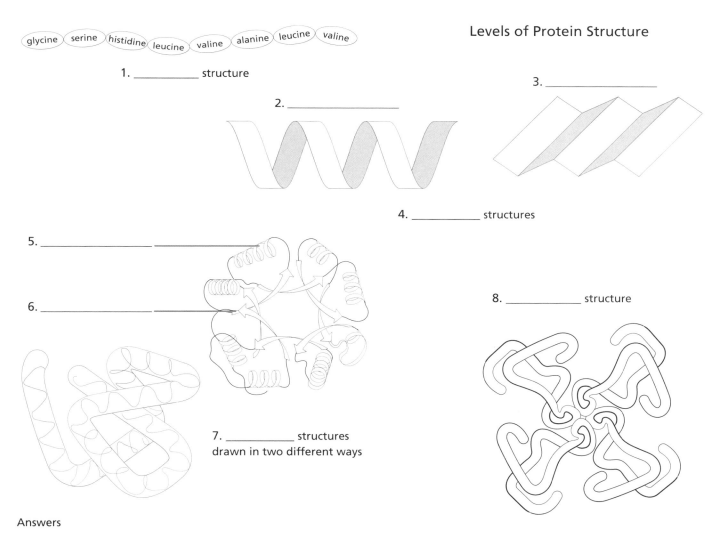

glycine — serine — histidine — leucine — valine — alanine — leucine — valine

1. _____ structure

2. _____

Levels of Protein Structure

3. _____

4. _____ structures

5. _____ _____

6. _____

7. _____ structures
drawn in two different ways

8. _____ structure

Answers

1. primary, 2. α-helix, 3. β-strand, 4. secondary, 5. α-helix, 6. β-strand (pleated sheet), 7. tertiary, 8. quaternary

Nucleic Acids

Nucleic acids are the information molecules of cells: just as you can store information in the numeric code of bits on your computer, cells store information in the chemical code of nucleic acids. DNA makes up chromosomes and stores all the information needed for the structure and function of cells. When cells and organisms reproduce, they transmit information by giving their descendants a copy of their DNA. Ribonucleic acid (RNA) molecules perform many functions in cells. For example, messenger RNA (mRNA), transfer RNA (tRNA), and ribosomal RNA (rRNA) work together to construct proteins based on the information stored in DNA.

The monomer of nucleic acids is the nucleotide. Each nucleotide has three components: a 5-carbon sugar, a nitrogenous base, and a phosphate group. Nitrogenous bases are ring structures that contain nitrogen atoms. The bases adenine (A) and guanine (G) are both purines because they have two rings, while cytosine (C), thymine (T), and uracil (U) are pyrimidines, which have one ring. Nucleotides join together in polynucleotide chains by covalent bonds between the phosphate group of one nucleotide and the 3′ carbon in the sugar of another nucleotide. This creates an alternating pattern of sugars and phosphates along the backbone of the polynucleotide, while the nitrogenous bases extend off the backbone.

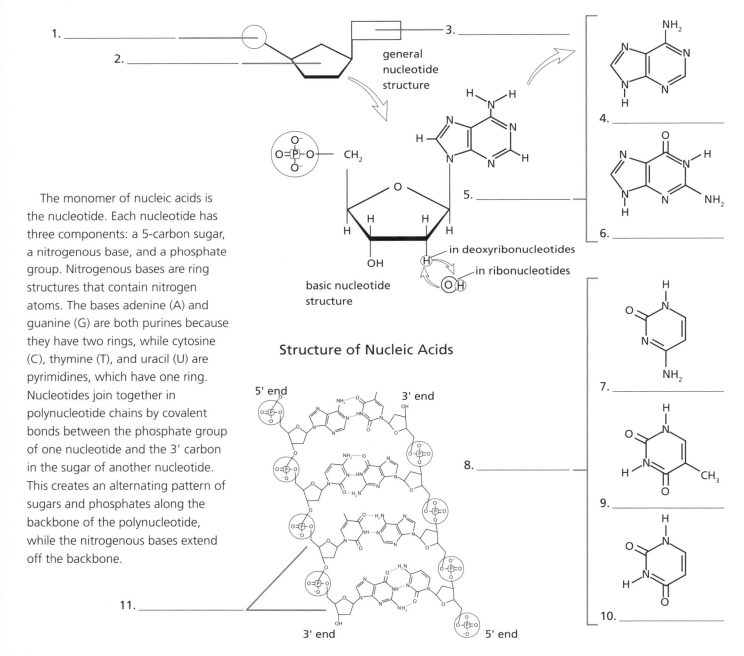

general nucleotide structure

basic nucleotide structure

in deoxyribonucleotides
in ribonucleotides

Structure of Nucleic Acids

5′ end   3′ end

3′ end   5′ end

Close-Up of DNA and RNA

DNA is famously known as the double helix because it's composed of two twisted polynucleotide strands. The two strands attach to each other by hydrogen bonds between their nitrogenous bases. You can think of the structure of DNA as a twisted ladder, where the sugar-phosphate backbones of the two strands are the side rails of the ladder, and the paired bases make up the rungs. DNA nucleotides contain the nitrogenous bases A, G, C, and T, which have a very specific relationship with one another. Based on their chemical structure, A bonds only with T, and G bonds only with C. Biologists refer to these partnerships as base pairs. The sugar in DNA is deoxyribose, which has one less oxygen atom than the sugar ribose.

RNA molecules are said to be single-stranded because they don't have a partner strand. Instead, many RNA molecules fold up and form internal hydrogen bonds that hold them in a three-dimensional shape. The bases in RNA are A, G, C, and U, and the sugar is ribose. As in DNA, the bases pair specifically according to their hydrogen bonding sites: A pairs with U and G pairs with C.

RNA

DNA

1. _____

2. _____

3. _____

4. _____

6. _____   5. _____

Structure of DNA and RNA

RNA

7. _____

8. _____

9. _____

10. _____

11. _____

DNA

12. _____

13. _____

Answers

The Prokaryotic Cell

All bacteria and archaea have cells with a prokaryotic structure. These cells have three major characteristics that distinguish them from eukaryotic cells (see p. 25): they're about ten times smaller than eukaryotic cells, they don't store their DNA in a membrane-bound nucleus, and they don't have membrane-bound organelles.

Prokaryotic cells share many common structures, but particular cells will have different combinations of these depending on the strain of the organism and the growth conditions. All prokaryotic cells—and, in fact, all cells—have an outer boundary called a plasma membrane or cell surface membrane. Some prokaryotes, including most bacteria, have an additional reinforcing layer called a cell wall on top of the plasma membrane.

On the exterior of the cell, you might also find a sticky layer called a glycocalyx, which prokaryotes use for attachment and defense; small protein cables called pili or fimbriae (singular pilus and fimbria), used for attachment and walking motility; and longer protein cables called flagella (singular: flagellum), used to propel prokaryotes through the environment. On the inside of the prokaryotic cell, you find the chromosome of the cell in a space called the nucleoid, along with smaller pieces of DNA called plasmids. You also find many of the tiny ribosomes that cells use to make proteins. Altogether, the thick fluid and internal parts of the cell make up the cytoplasm.

0.5–10 µm

1. _____

2. _____

3. _____

4. _____

5. _____

6. _____

7. _____

8. _____

Generalized Prokaryotic Cell

Answers

The Eukaryotic Cell

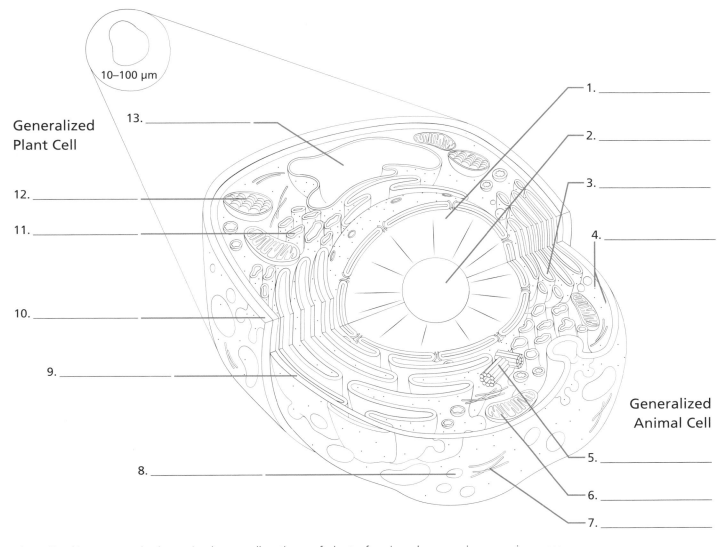

10–100 μm

Generalized Plant Cell

13. _____

12. _____

11. _____

10. _____

9. _____

8. _____

1. _____

2. _____

3. _____

4. _____

Generalized Animal Cell

5. _____

6. _____

7. _____

The cells of humans and other animals, as well as those of plants, fungi, and many microorganisms, are eukaryotic cells. The defining feature of eukaryotic cells is that their DNA is contained in a membrane-bound nucleus. In addition, eukaryotic cells have many internal membranes that form compartments within the cells, including additional membrane-bound organelles. Eukaryotic cells range in size (10–100 μm), but they are typically much larger than prokaryotic cells.

All eukaryotic cells also have a plasma membrane, ribosomes, and cytoplasm. Almost all have mitochondria—organelles that convert the energy in food to a form that cells can use (a molecule called adenosine triphosphate, or ATP). Almost all have a membranous system for making and moving molecules—the endomembrane system. This includes the rough endoplasmic reticulum (RER) for making proteins; the smooth endoplasmic reticulum (SER) for making lipids; the Golgi apparatus for modifying molecules and targeting them for their destination; and small spheres called vesicles, which help ship materials around the cell. Protein microtubules and filaments called the cytoskeleton help reinforce the membranes of the cell and provide tracks that aid the movement of vesicles.

Answers

1. nucleus, 2. nucleolus, 3. Golgi apparatus, 4. ribosomes, 5. centriole, 6. mitochondrion, 7. cytoskeleton, 8. vesicle, 9. plasma membrane, 10. cell wall, 11. rough endoplasmic reticulum, 12. chloroplast, 13. vacuole

Plasma Membrane

All cells have a membrane boundary that separates the cytoplasm of the cell from its environment. Proteins and lipids, such as phospholipids, each make up almost half of the membrane. The phospholipids form a bilayer, with their hydrophilic heads turned outward and their hydrophobic tails pointing inward. Integral proteins span the entire width of the membrane, while peripheral proteins associate with the edges. Carbohydrates attach to the outside of the cell to form glycolipids, and other lipids insert between the phospholipids. In animal cells like the one shown here, cholesterol is a significant component of the membrane.

The plasma membrane helps maintain homeostasis by controlling substances entering and exiting the cell. Although small nonpolar molecules can slip through, the hydrophobic center of the bilayer is an effective barrier to molecules that are either large or hydrophilic. These molecules require the help of integral proteins in order to cross the membrane. Channel proteins are integral proteins that have a tunnel through their center. Carrier proteins pick up molecules on one side of the membrane and then change their shape to move the molecule to the other side. Both types of proteins may be closed unless they get a signal that tells them to open. Carbohydrates are important to cellular signaling, and lipids like cholesterol play a role in membrane stability.

Cell

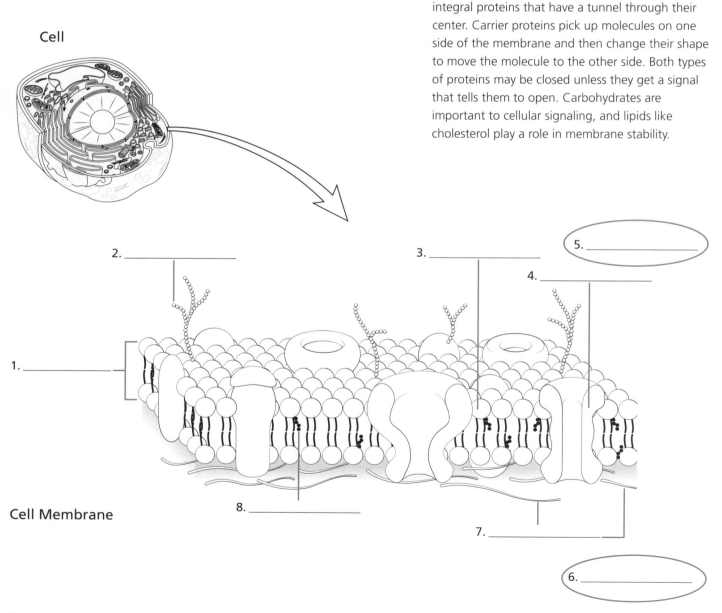

2. _____

3. _____

5. _____

4. _____

1. _____

Cell Membrane

8. _____

7. _____

6. _____

Answers

Membrane Transport

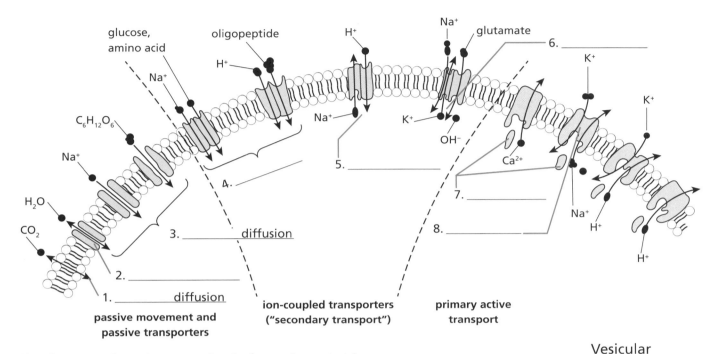

glucose, amino acid

oligopeptide

glutamate

$C_6H_{12}O_6$

H_2O

CO_2

6. _____

5. _____

4. _____

3. _____ diffusion

2. _____

1. _____ diffusion

7. _____

8. _____

passive movement and passive transporters

ion-coupled transporters ("secondary transport")

primary active transport

The plasma membrane is a very active site for moving materials into and out of the cell. The cell doesn't need to provide energy when molecules move passively by diffusion from areas where they are more concentrated to areas where they are less concentrated. However, if a cell wants to concentrate a molecule, it needs to provide energy for the active transport of that molecule from low to high concentration.

Diffusion happens in several ways. If a molecule is small and nonpolar, it may cross the phospholipid barrier directly by simple diffusion. Molecules that are either polar or larger may cross by facilitated diffusion with the help of a channel protein. Although water molecules are polar, they can pass through by simple diffusion called osmosis. Cells that move large amounts of water typically do so by facilitated diffusion using proteins called aquaporins.

Very large molecules may be moved across the membrane by endocytosis or exocytosis. During endocytosis, the cell membrane pinches together, forming a vesicle that brings molecules into the cell. The binding of a signaling molecule may trigger receptor-mediated endocytosis. During exocytosis, cellular materials travel to the cell membrane in a vesicle, which then fuses with the membrane, allowing the materials to be secreted outside the cell.

Vesicular Transport (Cytosis)

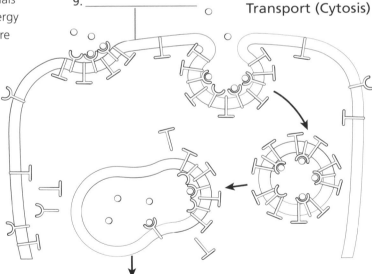

9. _____

10. _____

11. _____

12. _____

Answers

Close-Up of the Nucleus

The nucleus is defined by the nuclear envelope, a double membrane that separates the inside of the nucleus from the cytoplasm of the cell. The nucleus contains the genetic material of the cell in the form of DNA. Each strand of DNA winds around proteins to form a chromosome. When cells are not in the process of dividing, their chromosomes are spread loosely throughout the nucleus, forming a material called chromatin. When cells get ready to divide, the chromosomes coil up tightly and become visible as individual units.

Cells use the information encoded in their DNA to build molecules like RNAs and proteins. Ribosomal RNA (rRNA) molecules combine with proteins inside the nucleus to form the subunits of ribosomes. The assembly of the subunits occurs in dense regions of the nucleus called the nucleolus (plural: nucleoli), which is visible as a dark-staining region.

In order to do their jobs for the cell, ribosomal subunits and other types of RNA must move out of the nucleus and into the cytoplasm. Proteins called nuclear pore proteins assemble into structures that span the nuclear envelope, creating channels called nuclear pores that allow molecules to move between the nucleus and the cytoplasm.

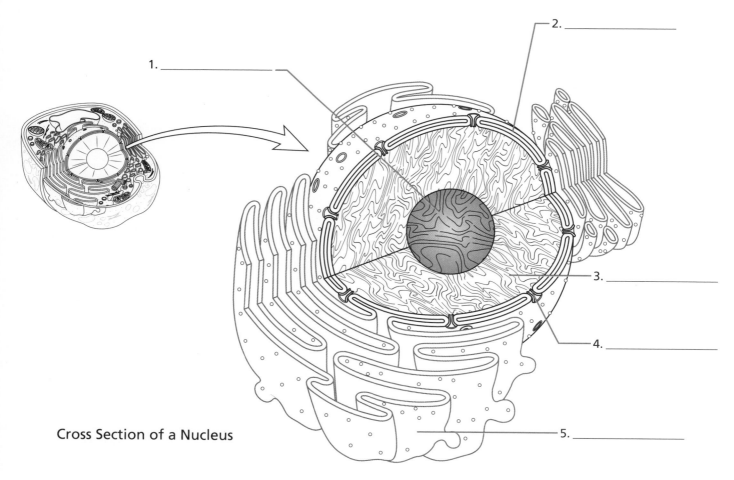

1. _____

2. _____

3. _____

4. _____

5. _____

Cross Section of a Nucleus

Answers

The Endomembrane System

The endomembrane system consists of the endoplasmic reticulum (ER), Golgi apparatus, transport vesicles, and lysosomes. These components work together to manufacture cellular components and ship them where they need to go, and also to break down and recycle old cellular components. Manufacturing begins at the ER, which extends from the nuclear envelope. Ribosomes attach to the rough ER (RER) where they synthesize proteins that will either become part of membranes or that will be secreted from the cell. The smooth ER (SER) is tubular and lacks ribosomes. It makes lipids for the cell, placing them into the membrane of the ER. The membrane of the ER can pinch off to form transport vesicles that carry the new proteins and lipids to their destinations.

Transport vesicles from the ER may travel to the Golgi apparatus, a stack of flattened membrane disks that help finish new membrane proteins. Vesicles from the ER fuse into the closest membranes of the Golgi (the *cis*-Golgi), depositing lipids and proteins into the Golgi membrane. Enzymes make changes to the proteins as they move through the stack of membranes, until they reach the side farthest from the ER (the *trans*-Golgi). The Golgi membrane then pinches off to form a new vesicle that is targeted to its final destination. If the vesicle contains hydrolytic enzymes, it may fuse with another vesicle to form a lysosome that can break down unwanted materials.

Cross Section of the Endomembrane

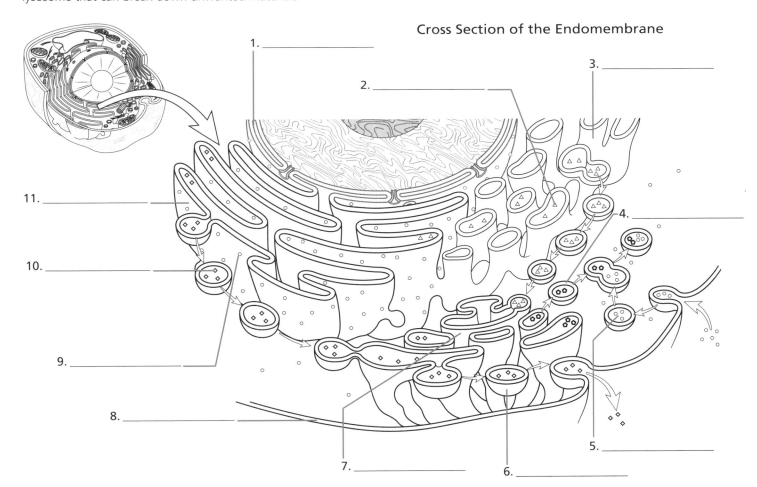

1. _____

2. _____

3. _____

4. _____

5. _____

6. _____

7. _____

8. _____

9. _____

10. _____

11. _____

Answers

Close-Up of the Mitochondrion

Just as a power plant provides usable energy to a community, so the mitochondrion provides usable energy to cells. It does this through the process of cellular respiration, a series of chemical reactions that make small changes to food molecules so that chemical energy from food can be transferred to the energy carrier, ATP. Cells use the energy from ATP to power endergonic processes (those requiring energy), such as biosynthesis, movement, and transport.

Eukaryotic cells gained mitochondria when an ancestor engulfed a free-living bacterium and the two cells developed or evolved a mutualistic form of symbiotic relationship, one that benefited both organisms. You can see evidence of this ancient event in the structure of the mitochondrion: it has two membranes—the outer membrane and the inner membrane—plus its own DNA and ribosomes, both of which reveal a close relationship to living bacteria.

Two different stages of cellular respiration take place in the mitochondrion. First, the Krebs cycle (or TCA cycle) occurs in the matrix. Then, oxidative phosphorylation uses an electron transport chain made of proteins in the inner membrane. The folds of the inner membrane, called cristae, increase the area available for this process. As the proteins pass electrons through the chain, some proteins actively transport protons (H$^+$) into the intermembrane space between the outer and inner membranes.

Cross Section of a Mitochondrion

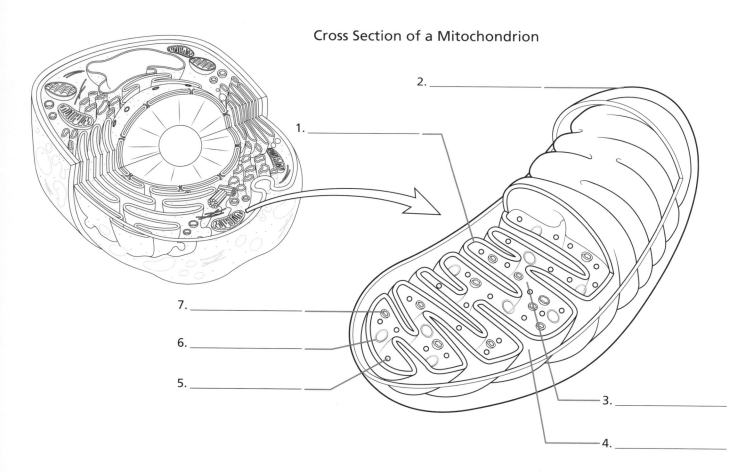

1. _____
2. _____
3. _____
4. _____
5. _____
6. _____
7. _____

Answers

1. inner membrane (cristae), 2. outer membrane, 3. matrix, 4. intermembrane space, 5. ribosome, 6. DNA, 7. granule

Close-Up of the Chloroplast

The chloroplast captures energy from the sun and uses it to make food molecules like glucose ($C_6H_{12}O_6$) from carbon dioxide (CO_2) and water (H_2O). Like mitochondria, chloroplasts resulted from an ancient symbiotic event between ancestors of the eukaryotic cell and free-living photosynthetic bacteria. The outer and inner membranes that form the envelope of the chloroplast probably resulted from ancestral eukaryotes engulfing the bacteria and bringing them into the cell in a vesicle. In addition to having two membranes, chloroplasts also contain prokaryotic ribosomes and DNA as a legacy of their former independence.

Photosynthesis consists of two main phases that each take place in a different part of the chloroplast. During the light reactions, which take place in the membranes of the thylakoid (a compartment within the chloroplast), chlorophyll and an electron transport chain work together to capture energy from the sun and transfer it to the energy carrier, ATP. At the same time, the chain takes electrons from water and passes them first to chlorophyll, then through the chain, and finally to the electron carrier, nicotinamide adenine dinucleotide phosphate (NADPH). During this process, some proteins actively transport protons (H^+) into the interior of the thylakoids.

The ATP and NADPH from the light reactions of photosynthesis move to the interior (stroma) of the chloroplast, where they participate in the Calvin cycle, a series of reactions that produce glucose from carbon dioxide. Plants combine glucose molecules into long chains of starch that may be stored as starch granules within the chloroplast.

Cross Section of a Chloroplast

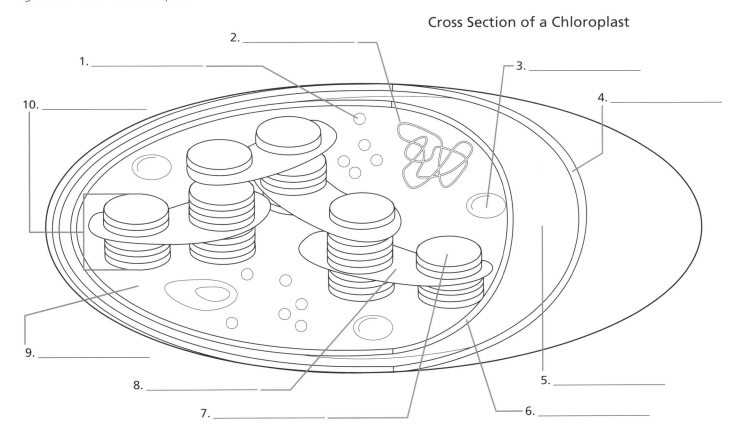

1. _____

2. _____

3. _____

4. _____

5. _____

6. _____

7. _____

8. _____

9. _____

10. _____

Answers

Close-Up of the Cytoskeleton

The cytoskeleton is a dynamic system of protein cables that both reinforces cellular structures and facilitates cellular motility. Although illustrations of cells often omit these proteins for simplicity, when scientists stain actual cells to show cytoskeletal proteins, they reveal a network of fibers that spreads through the cell. Scientists categorize cytoskeletal proteins in three groups based on the diameter of the protein.

Microfilaments, made of the protein actin and hence called actin filaments, are the thinnest of the cytoskeletal proteins (3–7 nm). They provide structural support to cellular extensions like the microvilli on epithelial cells and the filopodia of nerve cells. The assembly and disassembly of microfilaments extends the plasma membranes to form the pseudopods of amoebas and phagocytes. Actin works with the motor protein myosin to enable the contraction of muscle cells and the formation of the contractile ring during cell division.

Many different proteins make up intermediate filaments, which are about 10 nm in diameter. These proteins reinforce cellular structures like the nuclear envelope, the long axons of nerve cells, and the plasma membrane. Microtubules, made of the protein tubulin, are the thickest cytoskeletal proteins (20–25 nm). They form the mitotic spindle that moves chromosomes during cellular division and work with motor proteins to direct the beating of cilia and eukaryotic flagella. In order to move materials within the cell, motor proteins attach to vesicles and organelles, then "walk" along the microtubule tracks in the cell.

Structure and Arrangement of Cytoskeletal Proteins in a Cell

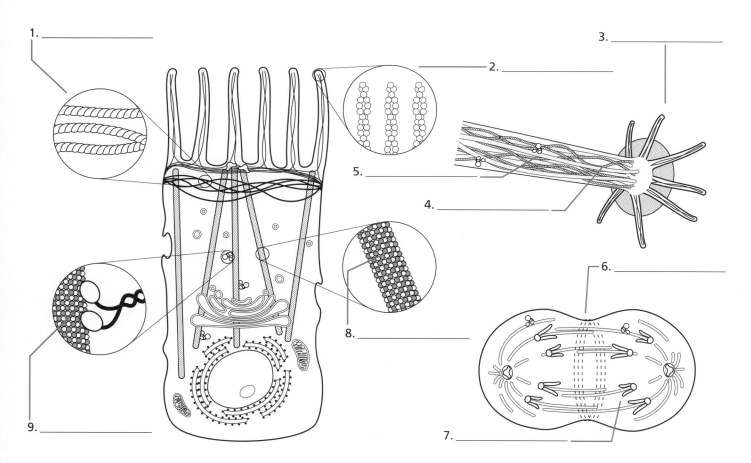

1. _____

2. _____

3. _____

4. _____

5. _____

6. _____

7. _____

8. _____

9. _____

Answers

Viral Structure

Viruses are tiny particles made primarily of nucleic acids surrounded by a protein coat called a capsid. They are so small (10–330 nm) that they can be seen only with an electron microscope. They aren't made of cells, so they lack the structures found in prokaryotic and eukaryotic cells; they don't have plasma membranes, ribosomes, or organelles. The reason viruses can reproduce without having the structures for protein synthesis or energy transfer is simple: they hijack cells, take them over, and use their materials and processes to make more of themselves. Some viruses always reproduce immediately after attacking a host cell, but some can enter a dormant phase. In bacteriophages, the viruses that attack bacteria, scientists call these two types of reproductive cycles the lytic and lysogenic cycles.

Viruses can be organized by their basic shapes. In symmetrical viruses, a polyhedral capsid surrounds the nucleic acid. In helical viruses, the capsid forms a long tubular shape. Some animal viruses cover themselves in viral envelopes that they make by modifying the host plasma membrane with viral proteins. These viruses may appear spherical because of the envelope, but a polyhedral or helical capsid may be present underneath. Some viruses have surface proteins called spikes that help them attach to the host cell. Viruses like the bacteriophages may have various protein appendages, resulting in a complex structure.

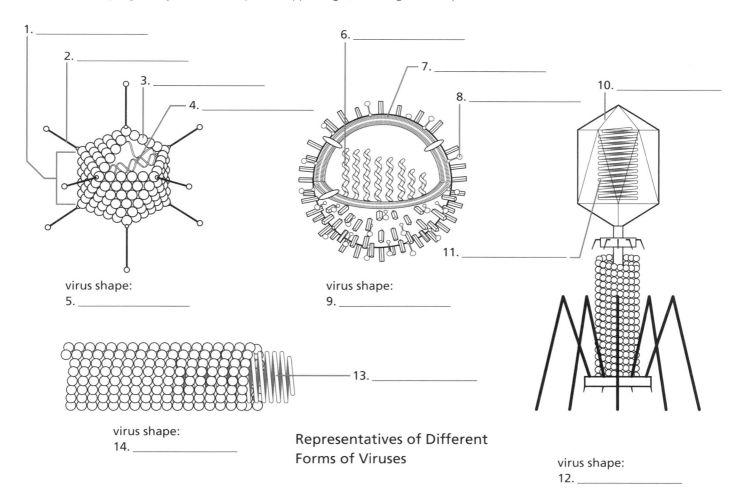

1. _____

2. _____

3. _____

4. _____

virus shape:
5. _____

6. _____

7. _____

8. _____

9. virus shape:
9. _____

11. _____

10. _____

12. virus shape:
12. _____

13. _____

virus shape:
14. _____

Representatives of Different Forms of Viruses

Answers

The Lytic Cycle of a Bacteriophage

During the lytic reproductive cycle of a bacteriophage, the virus attacks a bacterial cell, directs the cell to produce more viral particles, and then lyses, or bursts, the host cell so that the particles are released. Five essential steps occur during this cycle: attachment, penetration, biosynthesis, maturation, and release. The virus attaches to the host when one of its proteins binds to a host protein. Attachment is very specific: a virus can infect a cell only if it possesses the "key" with the right shape to pick the "lock" on the cell. The virus penetrates the host by inserting its genetic material into the cell. Bacteriophages often destroy the host DNA soon after penetration. Biosynthesis refers to the creation of new viral parts, including copies of the viral genetic material and production of capsid proteins. The viral genome directs construction, but the ribosomes, energy, and building materials come from the host cell. The new viral parts self-assemble into complete viral particles during maturation. Finally, the virus lyses the host cell during release, and the new viral particles burst out to seek new host cells.

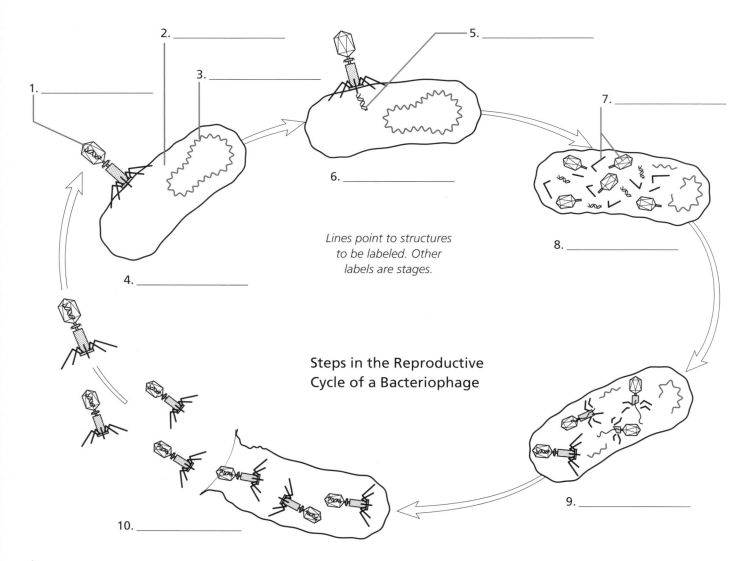

2. _____

5. _____

3. _____

1. _____

7. _____

6. _____

Lines point to structures to be labeled. Other labels are stages.

8. _____

4. _____

Steps in the Reproductive Cycle of a Bacteriophage

9. _____

10. _____

Answers

The Lysogenic Cycle of a Bacteriophage

Temperate bacteriophages can alternate between the lytic reproductive cycle and a lysogenic cycle during which they remain dormant inside a host cell for a period of time. The bacteriophages attach and penetrate the host cell just as they would during the lytic cycle. However, instead of immediately taking over the host cell and making more viral particles, they use genetic recombination to integrate a copy of their genetic material with the host DNA. The lysogenized host cell remains alive and fully functional and it may even be able to make new proteins due to the viral DNA. Every time the host cell divides, it copies the inactive viral DNA, or prophage, along with its own DNA, essentially generating a population of infected cells. Environmental changes can trigger changes in the host cell, leading to the release of the prophage from the bacterial DNA. The prophage becomes active and returns to the lytic phase, completing biosynthesis, maturation, and release.

Steps in the Lysogenic Cycle of a Bacteriophage

1. _____
2. _____
3. _____
4. _____
5. _____
6. _____
7. _____
8. _____
9. _____
10. _____
11. _____
12. _____

Lines point to structures to be labeled. Other labels are stages.

Answers

1. phage, 2. bacterial cell, 3. bacterial DNA, 4. attachment, 5. phage DNA, 6. penetration, 7. recombination (integration), 8. prophage, 9. cellular reproduction, 10. biosynthesis, 11. maturation (assembly), 12. release

The Reproductive Cycle of HIV

The human immunodeficiency virus (HIV) attaches to cells in the human immune system that have a surface protein called CD4. Proteins on the surface of the virus bind first to CD4, then to a second protein called a co-receptor. After it has attached, the virus penetrates the host cell by fusion: the viral envelope fuses with the host plasma membrane, and the viral RNA genome and some enzymes enter the host cell. The viral enzyme reverse transcriptase makes a DNA copy from the viral RNA. The viral DNA passes through the nuclear envelope and integrates into the host genome.

If conditions in the host cell favor reproduction of the virus, the viral DNA will direct synthesis of new viral RNA and proteins. Some viral proteins travel to the host plasma membrane, changing it into a new viral envelope. The viral genome and other proteins assemble near the modified plasma membrane and begin to push out of the cell by budding so that the envelope wraps around the viral capsid. Viral enzymes finish the process of maturation as the viral particles exit the cell. Even if the virus becomes latent, the host cell will copy the viral genetic material, along with its own, every time the cell divides.

Steps in Infection of a Cell by the HIV Virus

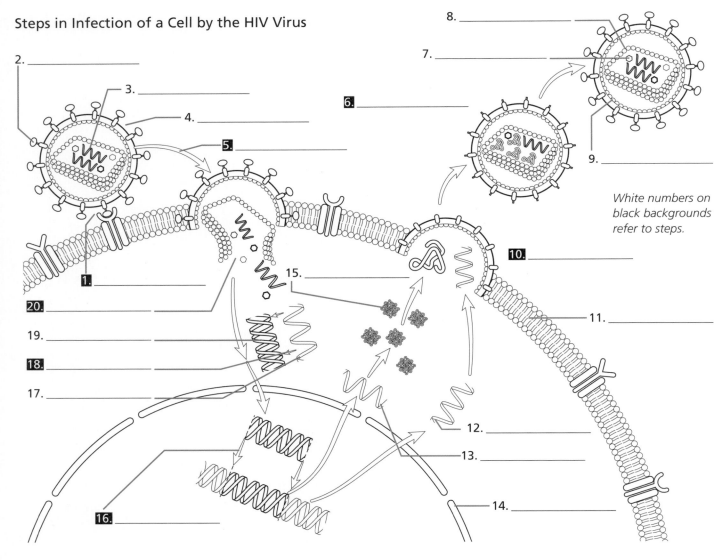

White numbers on black backgrounds refer to steps.

Answers

Cellular Energy

To stay alive, all living things need energy to maintain their organization and homeostasis. Some organisms can capture energy from environmental sources and convert it into chemical energy, which they then store in carbohydrate molecules. Scientists call these types of organisms autotrophs, which means "self-feeding." Plants are autotrophs that make carbohydrates through the process of photosynthesis: they use light energy to rearrange the atoms in carbon dioxide and water molecules, producing carbohydrates such as the sugar glucose. In eukaryotes, photosynthesis takes place inside chloroplasts.

White numbers on black backgrounds refer to processes.

Overall Energy Cycle

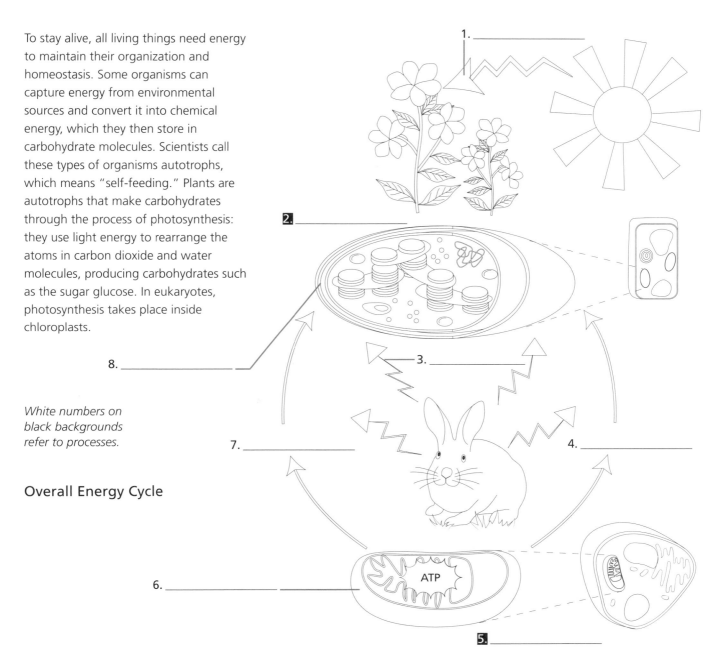

Living things that can't make their own food rely on organic molecules made by autotrophs. These types of organisms are called heterotrophs, which means "other-feeding," because they have to eat other organisms to get energy. Heterotrophs may feed on autotrophs such as plants directly, or they may eat other heterotrophs. Heterotrophs and autotrophs both transfer energy from food molecules into an energy carrier called adenosine triphosphate (ATP). Eukaryotes transfer energy from food to ATP by a process called cellular respiration, most of which takes place in the mitochondria. Once cells transfer energy to ATP, they can use the ATP to transfer energy into energy-requiring processes like biosynthesis (building molecules), transport, movement, repair, and reproduction.

Answers

1. light energy, 2. photosynthesis, 3. heat energy, 4. glucose and oxygen, 5. cellular respiration, 6. mitochondrion, 7. carbon dioxide and water, 8. chloroplast

The ADP/ATP Cycle

Cells use ATP to provide energy for endergonic (energy-requiring) processes. Like all molecules, ATP stores chemical energy through the arrangement of its atoms. ATP is a useful energy carrier because of its three negatively charged phosphate groups, which are held together in a row by covalent bonds. Like charges repel like charges, so the arrangement of these three groups is a source of potential energy. Cells obtain energy from ATP by transferring one of its phosphate groups to another molecule, making energy available for endergonic processes. The remainder of the ATP molecule, called adenosine diphosphate, or ADP, has less stored energy than ATP.

Just as removing a phosphate from ATP makes energy available to the cell, attaching a phosphate to ADP (phosphorylation) requires energy from the cell. In order to maintain their pool of available ATP, cells must transfer energy from another source so that they can phosphorylate the ADP. Cells transfer energy from exergonic (energy-releasing) processes, like the breakdown of food, in order to produce ATP from ADP and inorganic phosphate. Overall, there's a constant energy cycle in cells; cells do cellular work, such as biosynthesis, movement, and repair, by transferring energy from ATP to these endergonic processes, and then they transfer energy from exergonic processes to regenerate ATP from ADP and phosphate.

Transfer of Energy to and from ATP

1. _____

2. _____

3. _____

4. _____

5. _____

6. _____

energy from
10. _____

energy available for
7. _____

9. _____

8. _____

Answers

Redox Reactions

"Redox" is scientist slang for oxidation-reduction reactions. Redox reactions involve the transfer of electrons from one molecule to another. Scientists use the term oxidation to refer to a molecule losing its electrons and reduction to refer to a molecule gaining electrons. (At first glance, it doesn't make sense for the term reduction to refer to a gain, but there are lots of mnemonics to remember this, such as OIL RIG: Oxidation Is Losing, Reduction Is Gaining.) Scientists use the term redox reactions because, in cells, oxidation and reduction always happen together: if one molecule loses electrons, those electrons move to another molecule.

A Redox Reaction Involving NAD⁺/NADH + H⁺

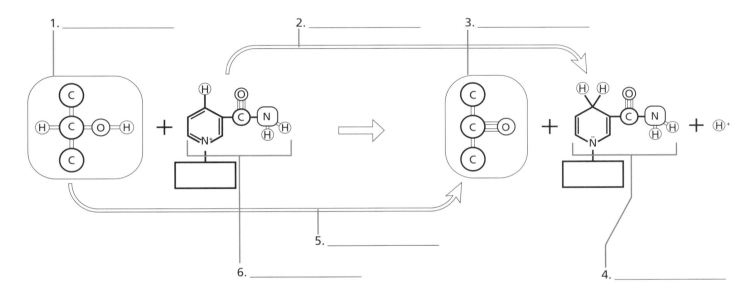

Cells often use electron-carrier molecules to facilitate the movement of electrons from one set of reactions to another. One of the most important electron carriers in cells is nicotinamide adenine dinucleotide. This molecule shifts between two forms depending on whether it's oxidized or reduced: NAD^+ refers to the oxidized form, while NADH refers to the reduced form. Cells oxidize organic molecules by transferring a pair of hydrogen atoms (each hydrogen atom has one electron and one proton) to NAD^+. NAD^+ accepts both electrons and one proton from the pair, converting to the reduced form NADH + H⁺ (the H⁺ here refers to the proton that's not accepted). NADH can now travel within the cell and provide electrons to a different chemical reaction. Cells constantly transfer electrons during reactions, which creates a cycle that converts NAD^+ to NADH + H⁺ and back again.

Answers

Metabolic Pathways

Metabolism refers to all the chemical reactions performed by a cell. Chemical reactions rearrange the structure of molecules, resulting in new molecules and changes in energy. In a single reaction, the molecules that exist before the reaction are the reactants; the molecules produced are the products. Cells conduct thousands of chemical reactions at any given moment, and the product of one reaction may be the reactant for another. Biologists show this by organizing related reactions into chains of reactions they call metabolic pathways. The first molecule in the chain is the reactant, the last molecule is the product, and all the molecules in between are intermediates. Metabolic pathways may be linear, or a single intermediate may act as the starting point for two different pathways, creating a branched pathway. Sometimes, metabolic pathways re-create the starting molecule, forming a cyclic pathway.

In order to sustain life, cellular metabolism must happen rapidly, and it must be carefully controlled. These requirements are met by enzymes, proteins that act as catalysts to speed up chemical reactions. Every reaction in a metabolic pathway requires its own unique enzyme to make that reaction happen. Because reactions can't happen without their enzymes, cells control metabolic pathways by controlling key enzymes in the pathway. When describing enzyme-catalyzed reactions, scientists refer to reactants as substrates.

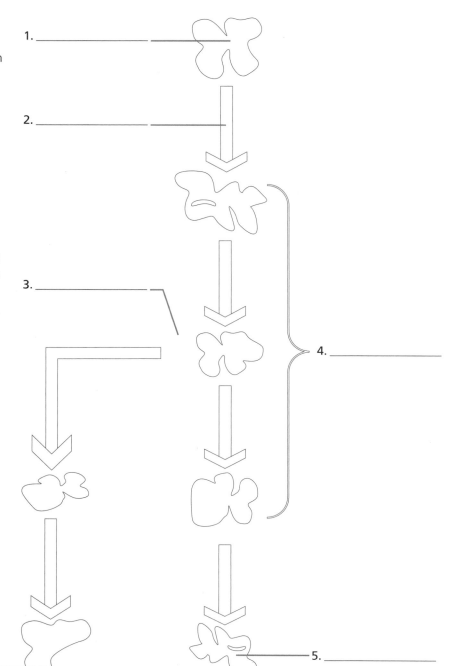

1. _____ _____

2. _____ _____

3. _____

4. _____

5. _____

6. _____ _____

Basic Features of Metabolic Pathways

Answers

Enzyme Structure

Most enzymes are proteins. The polypeptide chain that forms the protein folds up to create a unique shape for each enzyme. Part of the shape includes pockets that act as binding sites for other molecules. The active site is the binding site for the substrates of the reaction that the enzyme catalyzes. Allosteric sites are other sites for the binding of regulatory molecules.

The active site of an enzyme helps create conditions necessary for the chemical reaction to proceed. When the substrate binds to the active site, the enzyme shifts a little to create a perfect fit called an induced fit. Functional groups from the polypeptide chains of the enzyme line the active site. These groups can interact with the chemical groups on the substrate, making it more likely for the reaction to take place. After the reaction, the enzyme releases the products and returns to its original state. Overall, enzymes aren't used up or permanently changed by the reaction, and they can repeat catalysis over and over again.

Some enzymes require partner molecules, called cofactors and coenzymes, in order to catalyze a reaction. Cofactors are minerals, while coenzymes are organic molecules. For these enzymes, scientists call the polypeptide alone the apoenzyme. When the necessary cofactor or coenzyme is attached to create the functional enzyme, they call this molecule the holoenzyme.

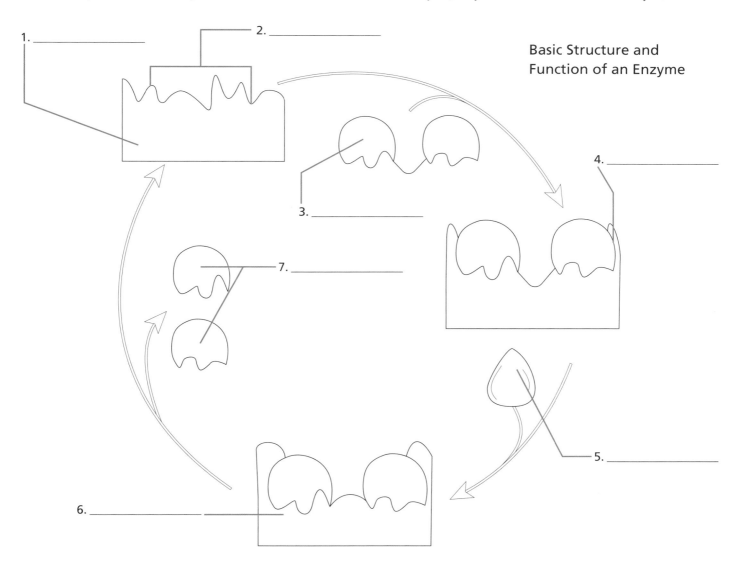

1. _____

2. _____

3. _____

4. _____

5. _____

6. _____

7. _____

Basic Structure and Function of an Enzyme

Answers

1. enzyme, 2. active site, 3. substrate, 4. induced fit, 5. water (H_2O), 6. hydrolysis, 7. products

Enzyme Regulation

Cells control their metabolism by regulating their enzymes. One way to regulate enzymes is through their allosteric sites. When molecules bind to the allosteric sites, the enzymes change their shape slightly, altering their active sites. When an allosteric activator binds, the active site changes in a way that improves the enzyme's ability to catalyze the reaction. An allosteric inhibitor has the opposite effect: it causes the active site to change so that it no longer binds to the substrate, effectively blocking the chemical reaction. Allosteric inhibition of enzymes is also called noncompetitive inhibition because the regulatory molecules don't compete with the substrate for the active site.

    End-product inhibition, also called feedback inhibition, is a simple and practical type of allosteric regulation. Cells use this type of regulation to control many important metabolic pathways. The final product of the pathway, the end product, acts as an allosteric inhibitor of a key enzyme in the pathway. This enzyme usually catalyzes one of the early reactions in the pathway and is sometimes involved in a reaction at a major branch point in a pathway. Thus, if the cell has plenty of the end product of a pathway, then the end product itself will act to shut down the pathway so that the cell can divert its resources to other needs.

Inhibition and Activation of Allosteric Enzymes

End-product Inhibition

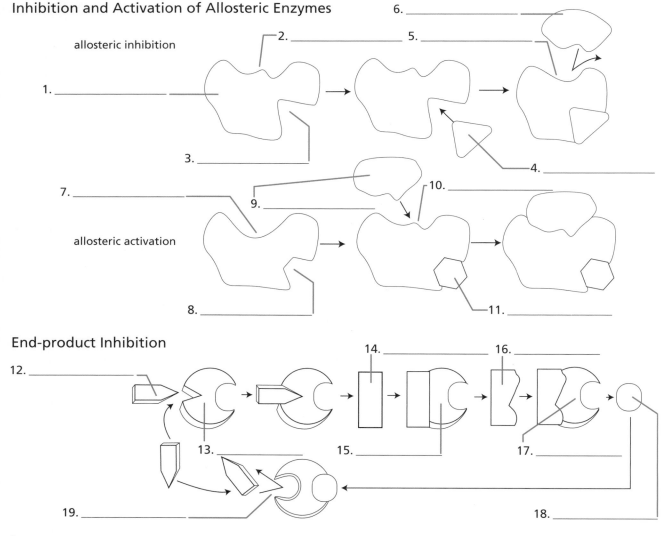

Answers

Cellular Respiration

Cellular respiration is essential for many living things on Earth because it's the metabolic pathway they use to transfer energy from their food to the energy carrier ATP. All types of food molecules can enter this pathway at various points, but biologists usually study the pathway beginning with glucose. During cellular respiration, cells oxidize glucose and ultimately transfer the electrons to oxygen gas, reducing the oxygen to water. The oxidation of glucose also results in the production of carbon dioxide molecules. Although the pathway consists of many small reactions, the entire process can be summarized as follows:

$$C_6H_{12}O_6 + 6O_2 \text{---}> 6CO_2 + 6H_2O$$

Biologists separate cellular respiration into three stages: glycolysis, the Krebs cycle, and oxidative phosphorylation. In eukaryotes, glycolysis occurs in the cytoplasm of the cell, while the Krebs cycle and oxidative phosphorylation occur in the mitochondrion. During glycolysis and the Krebs cycle, some reactions transfer energy to create ATP, while others transfer electrons to electron carriers such as NAD^+ and flavin adenine dinucleotide (FAD). During oxidative phosphorylation, cells transfer the electrons to an electron transport chain, resulting in the transfer of energy from the original food molecules to ATP for storage. If a single glucose molecule is fully oxidized by cellular respiration, the cell gains between 36 and 38 ATP molecules.

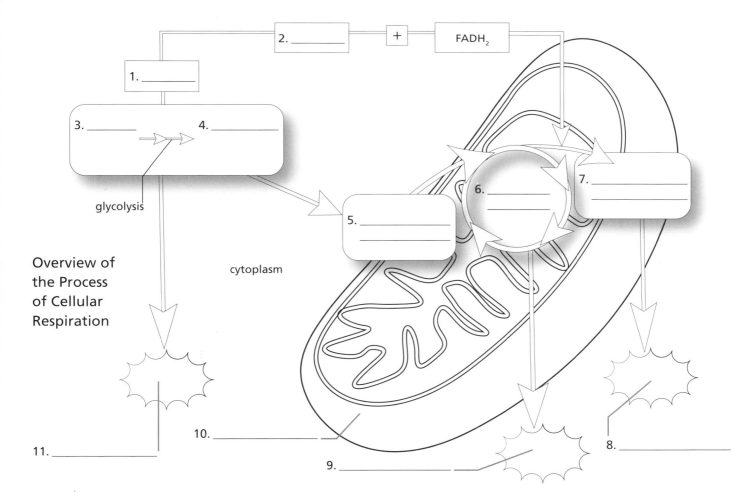

2. _____   +   FADH$_2$

1. _____

3. _____   4. _____

glycolysis

Overview of the Process of Cellular Respiration

cytoplasm

5. _____

6. _____

7. _____

10. _____

11. _____

9. _____

8. _____

Answers

Close-Up of Glycolysis

Glycolysis literally means the lysis, or breakdown, of glucose. The ten reactions of glycolysis ultimately break down glucose, which has six carbons, into two molecules of pyruvate, which has three carbons. During the process, cells store energy in ATP and electrons in NADH. The first half of glycolysis requires the input of energy from two molecules of ATP, but the second half yields enough energy to create four molecules of ATP, resulting in a net gain of two ATP molecules for the cell.

The first part of glycolysis rearranges the atoms in glucose and transfers energy from ATP into the molecules. First, cells transfer a phosphate from ATP to glucose, producing glucose-6-phosphate. The second reaction rearranges the atoms to produce fructose-6-phosphate. The third reaction phosphorylates the fructose-6-phosphate into fructose-1,6-bisphosphate. This intermediate has two phosphate groups and more stored energy than the original glucose.

The second part of glycolysis breaks the carbon backbone into two three-carbon molecules and transfers energy and electrons to form ATP and NADH. First, cells split the fructose-1,6-bisphosphate into one molecule of glyceraldehyde-3-phosphate (G3P) and one molecule of dihydroxyacetone phosphate (DHAP). The second reaction rearranges the atoms in DHAP to produce another molecule of G3P. Next, cells oxidize the G3P molecules, transferring electrons to NAD^+ to form NADH + H^+. Cells use some of the energy from this reaction to transfer inorganic phosphate to the intermediates, resulting in two molecules of 1,3-bisphosphoglycerate. Cells harvest energy by transferring a phosphate group from each intermediate to ADP, yielding 3-phosphoglycerate, then they rearrange the atoms in the intermediate to produce 2-phosphoglycerate. Cells remove a water molecule from this intermediate to produce phosphoenolpyruvate (PEP). In the final reaction, cells transfer a phosphate from each PEP to ADP, forming ATP and producing two molecules of the final product, pyruvate. For each molecule of glucose broken down through glycolysis, cells gain two molecules of NADH, two molecules of ATP, and two molecules of pyruvate.

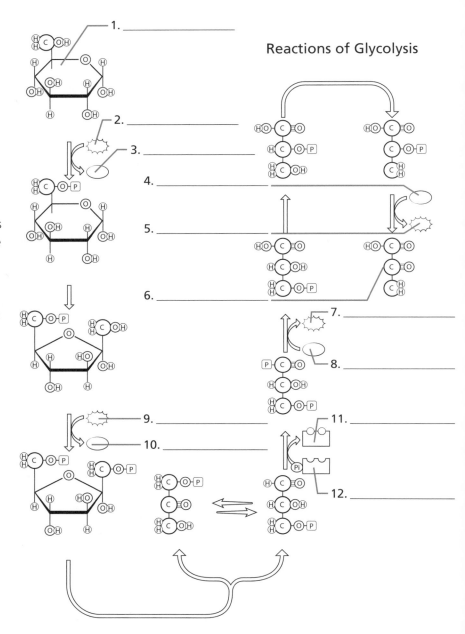

Reactions of Glycolysis

1. _____
2. _____
3. _____
4. _____
5. _____
6. _____
7. _____
8. _____
9. _____
10. _____
11. _____
12. _____

Answers

Close-Up of the Krebs Cycle

The Krebs cycle continues the oxidation of carbohydrate molecules to yield more ATP and reduced electron carriers for the cell. Before the Krebs cycle begins, cells oxidize pyruvate, transferring electrons to NAD^+. These reactions also remove carbon from an intermediate (decarboxylation), releasing a molecule of carbon dioxide, and attach coenzyme-A (CoA) to the carbon backbone. Ultimately, the reactions of pyruvate oxidation convert pyruvate (three carbon atoms) into acetyl-CoA (two carbon atoms), which can now enter the Krebs cycle.

The main function of the Krebs cycle is to oxidize food molecules so that energy and electrons can be captured for future use by the cell. The first reaction combines acetyl-CoA with oxaloacetate (four carbons), producing citrate (six carbons) and releasing CoA. The next reaction rearranges the atoms in citrate to produce isocitrate. The oxidation and decarboxylation of citrate produces NADH, carbon dioxide, and α-ketoglutarate (five carbons). Another oxidation and decarboxylation follows, producing NADH, carbon dioxide, and succinyl-CoA (four carbons). The next stage transfers energy, first to the energy carrier guanosine triphosphate (GTP) and then to ATP. This reaction produces succinate (four carbons) and releases CoA. Oxidation of succinate produces fumarate (four carbons) and transfers electrons to FAD. The addition of a water molecule converts fumarate to malate (four carbons). One final oxidation converts malate back to oxaloacetate, the starting molecule of the Krebs cycle, and transfers electrons to NAD^+.

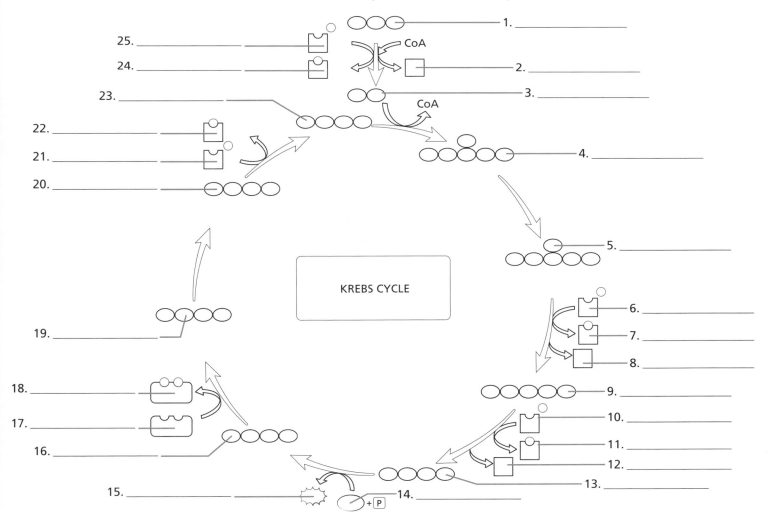

Answers

Close-Up of Oxidative Phosphorylation

Oxidative phosphorylation is a process that many cells use to produce ATP. It requires a group of specialized membrane proteins that participate in redox reactions to move electrons from one protein to the next, forming an electron transport chain. NADH and FADH$_2$ provide electrons to the chain. During aerobic respiration, the last protein in the chain transfers the electrons to oxygen gas. While the proteins in the electron transport chain move electrons, they also transfer energy, ultimately making it possible for cells to produce ATP.

Energy from the movement of electrons allows the proteins to transport protons (H$^+$) actively from one side of a membrane to the other, creating a concentration of protons. Just like water stored behind a dam, protons concentrated on one side of a membrane represent a source of potential energy to a cell. When cells allow protons to diffuse back across the membrane, they can harness that flow to use for cellular work, such as making ATP.

Oxidative phosphorylation relies on a protein called ATP synthase, which can synthesize ATP molecules from ADP and phosphate. ATP synthase contains a proton channel that allows protons to diffuse across the membrane. The diffusion of the protons (called chemiosmosis) causes a subunit of the protein to spin, which in turn allows the protein to transfer energy into the production of ATP.

Cellular Respiration in a Mitochondrion

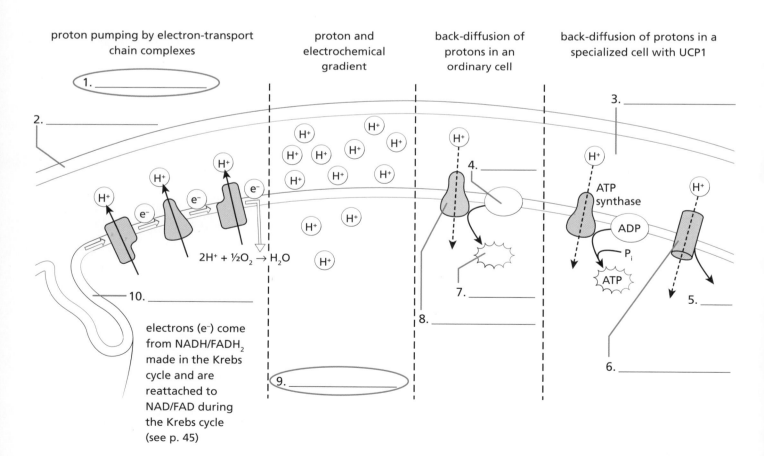

Answers

Fermentation

Fermentation allows cells to make ATP without an electron transport chain. Glycolysis produces two ATP molecules for every glucose molecule, so cells can get enough energy by repeating this process. However, every time cells carry out glycolysis, they transfer electrons to NAD^+, converting it to NADH. In order to repeat glycolysis, cells must transfer electrons from NADH to another molecule in order to recreate NAD^+. When cells run electron transport chains, they continuously recycle NADH. If cells lack oxygen or don't have an electron transport chain, however, they can use fermentation to meet their energy needs. Because fermentation doesn't require oxygen, it's an anaerobic process.

   Fermentation is glycolysis plus a recycling step that scientists call the fermentation step. The simplest fermentation of all is lactic acid fermentation. After glycolysis, cells transfer electrons from NADH to pyruvate, oxidizing NADH to NAD^+ and reducing pyruvate to lactic acid. Human muscle cells and the lactic acid bacteria you find in dairy products all use lactic acid fermentation.

   Ethanol fermentation is slightly more complicated. After glycolysis, cells decarboxylate pyruvate to produce carbon dioxide and acetaldehyde. Then cells transfer electrons from NADH to acetaldehyde, oxidizing NADH to NAD^+ and reducing acetaldehyde to ethanol. Yeasts perform this type of fermentation, which is why beer and wine contain ethanol. The carbon dioxide produced from the decarboxylation creates bubbles in beer and makes bread dough rise.

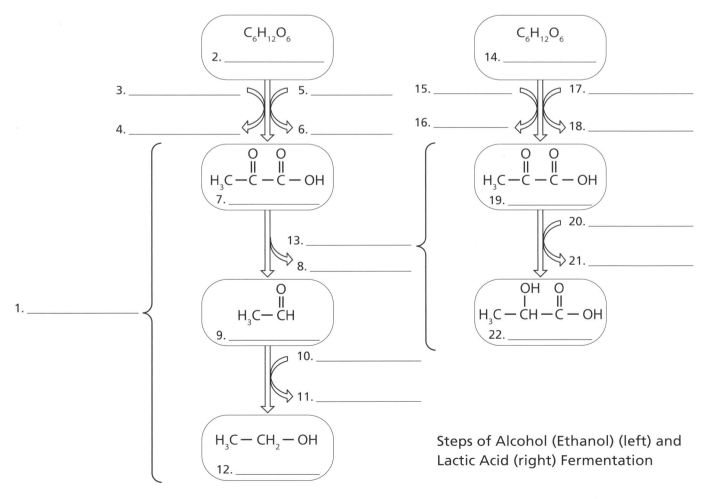

Steps of Alcohol (Ethanol) (left) and Lactic Acid (right) Fermentation

Answers

Photosynthesis

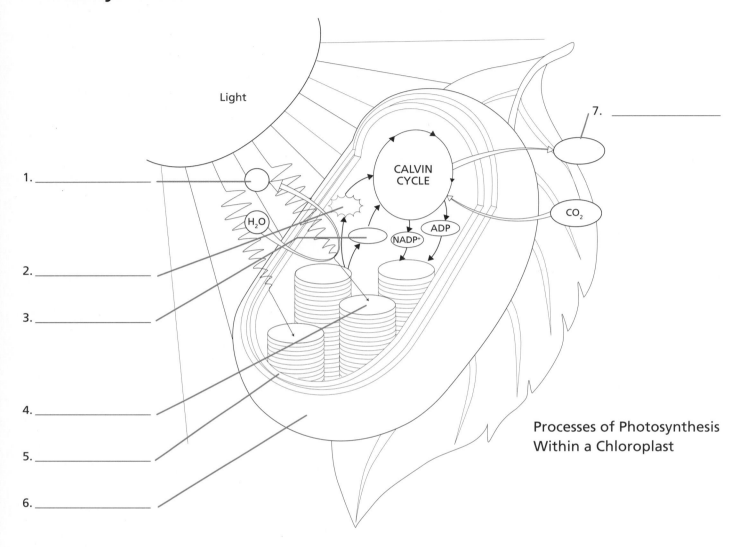

Light

1. _____

2. _____

3. _____

4. _____

5. _____

6. _____

7. _____

H₂O

CALVIN CYCLE

NADP⁺   ADP

CO₂

Processes of Photosynthesis Within a Chloroplast

The purpose of photosynthesis is to use the energy of sunlight to rearrange inorganic matter from the environment into organic matter in the form of sugars. The process can be broken down into two main parts: the light-dependent reactions, which capture the energy from the sun, and the light-independent reactions (called the Calvin cycle), which use the captured energy to rearrange carbon dioxide (CO_2) and water (H_2O) into sugars like glucose ($C_6H_{12}O_6$).

During the light-dependent reactions, light energy is captured by pigments in membranes called thylakoids, which are arranged into stacks called grana within the chloroplast. This energy helps the chloroplast split water molecules and is then transferred to the energy-rich molecule ATP. A molecule called nicotinamide adenine dinucleotide phosphate ($NADP^+$), in its reduced form NADPH, carries the hydrogen atoms from the water molecules, and the oxygen atoms exit the chloroplast and enter the environment as waste.

During the light-independent reactions, chloroplasts use the energy stored in ATP to rearrange CO_2 molecules captured from the environment. NADPH donates hydrogen atoms, reducing the carbon-containing molecules into sugars, which ultimately become the source of the energy and building materials used by almost all life on Earth. Photosynthetic organisms like plants use the sugars they make for themselves; other organisms must either eat the plants or eat something that ate a plant.

Answers

Close-Up of the Light Reactions

During the light reactions (photophosphorylation) taking place in photosynthesis, cells transform light energy into chemical energy stored in ATP. They also transfer electrons from an inorganic molecule such as water to the electron carrier NADPH.

Photosynthetic cells utilize an electron transport chain to transform light energy into chemical energy. Light-absorbing pigments, such as chlorophyll, are embedded in the same membrane as the chain. Chlorophyll absorbs light energy, which excites its electrons and allows them to transfer to proteins in the chain. The light energy also allows cells to split molecules like water and take electrons to replace the ones lost from chlorophyll. The splitting of water produces oxygen gas as waste.

The electron transport chain of the light reactions functions very similarly to that of oxidative phosphorylation. Proteins participate in redox reactions to move electrons down the chain until they reach the final electron acceptor, NADPH. The electron flow allows the active transport of protons across a membrane, which then diffuse back through ATP synthase so that the cell can make ATP.

In eukaryotes, the light reactions occur in chloroplasts, within the membranes of thylakoids. The chain captures light energy at two points: when photosystem II absorbs energy at the beginning of the chain and when photosystem I absorbs energy in the middle of the chain. As the electrons flow, protons are moved into the middle of the thylakoids. The ATP and NADPH move out into the stroma of the chloroplast, where they participate in the light-independent reactions (see p. 50).

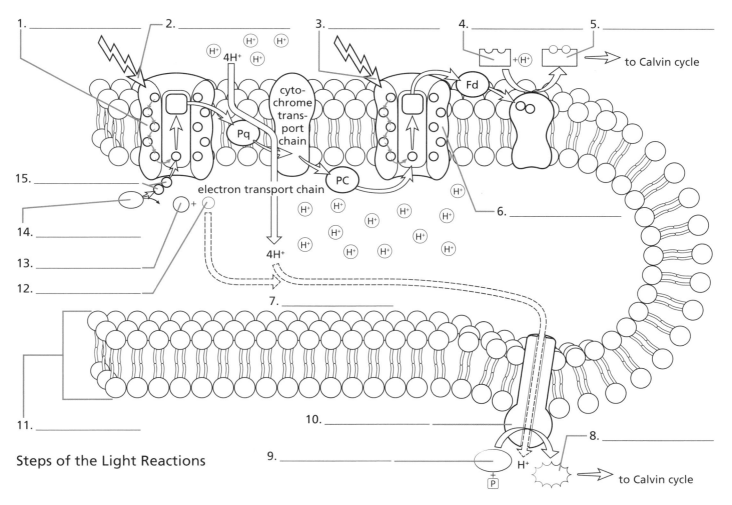

Steps of the Light Reactions

Close-Up of the Light-Independent Reactions

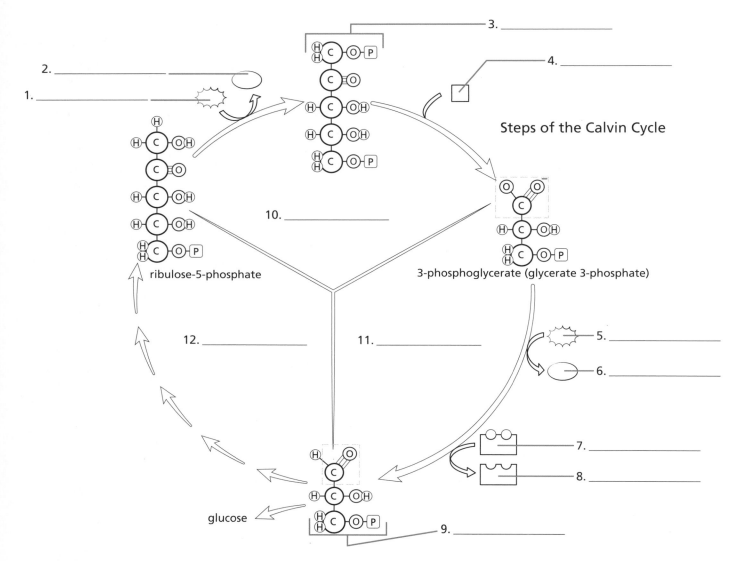

2. _____

1. _____

3. _____

4. _____

Steps of the Calvin Cycle

ribulose-5-phosphate

3-phosphoglycerate (glycerate 3-phosphate)

10. _____

12. _____

11. _____

5. _____

6. _____

7. _____

8. _____

glucose

9. _____

During the light-independent reactions (Calvin cycle), cells use energy from ATP and electrons from NADPH to reduce carbon dioxide to sugars such as glucose. In eukaryotes, this process occurs in the stroma of the chloroplast. The first step of this cycle is the capture, or fixation, of carbon dioxide from the environment. The enzyme ribulose-1,5-bisphosphate carboxylase/oxygenase (RuBisCO) binds carbon dioxide and transfers it to the sugar ribulose-1,5-bisphosphate (five carbons). The addition of carbon dioxide to this sugar produces an unstable intermediate (six carbons), which immediately breaks down into two molecules of 3-phosphoglycerate (PGA). The cell reduces the PGA to G3P using electrons carried by NADPH and energy from ATP. Cells can combine G3P molecules to produce glucose and combine glucose to produce starch for long-term storage.

To reset the cycle so that it can repeat, cells rearrange atoms in some of the G3P molecules to recreate the starting sugar ribulose-1,5-bisphosphate. This requires the input of energy from ATP. During the reduction step of the Calvin cycle, NADPH and ATP recycle back to NADP+ and ADP. These molecules can now be reused in the light reactions. Although the light-independent reactions don't require light directly, they do require the products of the light reactions (NADPH and ATP). So, when light isn't available to photosynthetic cells, this process shuts down along with the light reactions.

Answers

Cellular Connections

The cells of multicellular organisms attach to each other to form tissues and organs. Some cells connect to each other indirectly by sticky extracellular carbohydrates, much like bricks can be connected by sticky mortar. In plants, the primary cell walls of adjacent cells connect by a layer called the middle lamella. Animal cells produce a sticky matrix called the extracellular matrix (ECM) and may also be connected directly via proteins in their cell membranes.

Epithelia are important barrier tissues in animals, in which the cells are essentially stitched together by lines of proteins that connect the plasma membranes of adjacent cells. Tight junctions bring the cells very close together, forming a watertight seal that prevents materials from getting through the epithelia. Tight junctions aren't particularly strong, though, so many epithelial cells and muscle cells are attached to their neighbors by rivet-like desmosomes. Desmosomes consist of anchoring proteins under the plasma membrane of each cell that connect both to cytoskeletal proteins inside of the cells and to membrane proteins connecting the outer surfaces of adjacent cells.

Some cellular connections allow direct communication from the cytoplasm of one cell to the next. In plants, gaps in the cell wall allow the formation of membrane-lined channels called plasmodesmata. There may be 100–100,000 between adjacent cells. In animals, proteins may form rings in the membranes of adjacent cells, creating protein channels called gap junctions that connect one cell directly to the next. Both plasmodesmata and gap junctions enable rapid communication between neighboring cells.

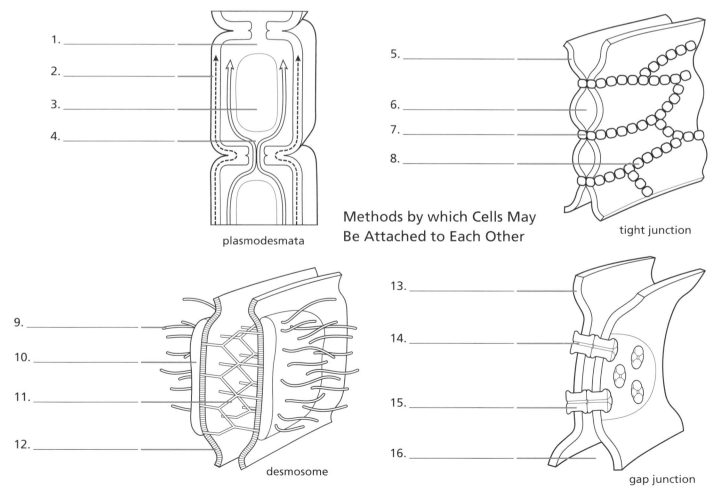

1. _____
2. _____
3. _____
4. _____

plasmodesmata

5. _____
6. _____
7. _____
8. _____

tight junction

Methods by which Cells May Be Attached to Each Other

9. _____
10. _____
11. _____
12. _____

desmosome

13. _____
14. _____
15. _____
16. _____

gap junction

Answers

1. cytoplasm, 2. cell wall, 3. vacuole, 4. plasmodesmata, 5. plasma membrane, 6. intercellular space, 7. tight junction, 8. junction proteins, 9. cytoskeletal filaments, 10. cytoplasmic plaque, 11. connecting filaments, 12. plasma membrane, 13. plasma membrane, 14. hydrophilic channel, 15. connexon, 16. intercellular space

Cell Signaling

Cells can communicate with distant cells via hormones or other signaling molecules, called ligands, which bind to receptors. Signaling molecules trigger many varied cellular responses via several different methods. Some ligands bind to ligand-gated ion channels, causing them to open or close. Other ligands use G proteins to pass their signal through the plasma membrane and on to second messengers (signal transduction).

In another type of signal transduction, the ligand binds to an enzyme-linked receptor, which spans the membrane and has an enzymatic portion on the inside of the cell. Ligand binding stimulates enzyme activity, resulting in the transfer of phosphate groups from molecules such as ATP and guanosine triphosphate (GTP) to the receptor and other proteins such as kinases. Phosphorylated kinases become active and, in turn, phosphorylate other kinases, causing a chain reaction that spreads the signal through the cell.

Finally, hydrophobic signaling molecules like steroid hormones may diffuse across the plasma membrane and bind to their receptors in the cytoplasm or nucleus of the cell. Ligand binding changes the conformation of the receptor, activating its ability to bind DNA. The ligand-receptor complex diffuses into the nucleus, where it causes a change in gene expression.

For labels 2, 4, 6, and 8, write what effect ligand binding has.

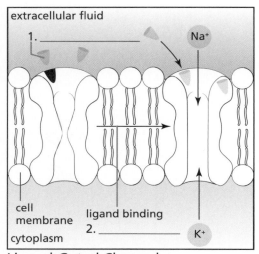

Ligand-Gated Channel

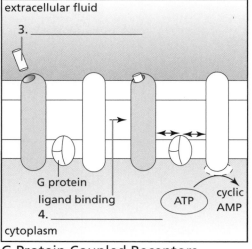

G Protein Coupled Receptors

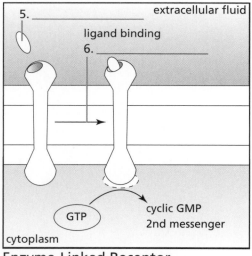

Enzyme-Linked Receptor

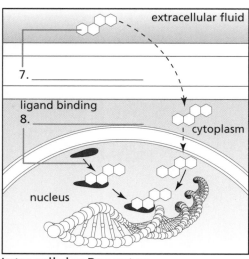

Intracellular Receptor

Answers

The Cell Cycle

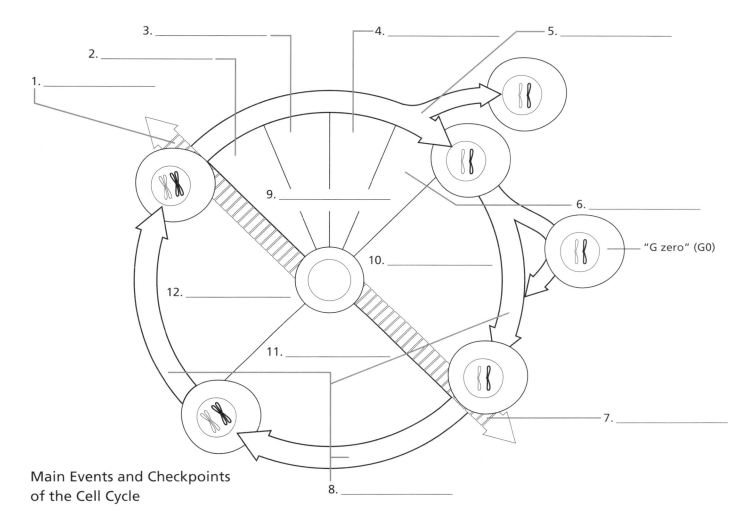

Main Events and Checkpoints of the Cell Cycle

3. _____
2. _____
1. _____
4. _____
5. _____
9. _____
6. _____
"G zero" (G0)
10. _____
12. _____
11. _____
7. _____
8. _____

Many eukaryotic cells have the ability to reproduce themselves via cell division. Scientists have mapped out a series of events, called the cell cycle, that can occur during the life of a cell as it alternates between dividing and nondividing states. Cells spend most of their time in interphase, the nondividing phase of the cell cycle, which is subdivided into gap 1 (G1), synthesis (S), and gap 2 (G2), plus a sort of idling phase called "G zero" (G0). Cells in G0 are alive and functioning but not growing or preparing to divide. Cells in G1 are growing, copying their organelles, and increasing in size. When they receive a signal to begin cell division, the cells enter S phase and make copies of all of their chromosomes. After DNA synthesis is complete, the cells enter G2 and check to make sure they copied the DNA correctly. From G2, cells enter M phase in order to divide by either mitosis or meiosis.

Cells use four checkpoints to ensure that cell division proceeds without error. Toward the end of G1, cells stop and check whether they're big enough, they have enough nutrients, their DNA is undamaged, and they've received a signal to divide. During the G2 checkpoint, proteins ensure that all the chromosomes were copied correctly and aren't damaged. Cells that enter M phase pass two more checkpoints: one in metaphase, which makes sure that all the chromosomes are attached to the spindle, and one in late anaphase, which checks whether all of the chromosomes separated correctly. Cells that can't pass these checkpoints may get stuck in one of the phases of the cell cycle and ultimately undergo apoptosis (programmed cell death).

Answers

1. G2 checkpoint, 2. prophase, 3. metaphase, 4. anaphase, 5. cytokinesis, 6. telophase, 7. G1 checkpoint, 8. mitosis (M phase), 9. interphase, 10. gap 1 (G1), 11. synthesis (S), 12. gap 2 (G2)

Asexual Reproduction

Asexual reproduction allows organisms to produce offspring that are genetically identical to the parent. One advantage to this method is that an organism doesn't have to locate a mate, which can be very valuable if individuals are dispersed widely. In addition, if an organism is successful in its current environment, it's likely that its identical offspring will also be successful.

Asexual reproduction is found in all types of living things, from bacteria, archaea, and protists to fungi, plants, and even animals. In bacteria and archaea, cells divide by a simple process called binary fission. Cells get larger, copy their single chromosomes, and split in two. Eukaryotic organisms use a slightly more complicated form of cell division called mitosis to ensure that their multiple chromosomes separate correctly.

Many familiar examples of asexual reproduction can be found in the world around you. If you use yeast to make beer or bread, the yeasts feed on the sugars and reproduce asexually in the food. Molds that grow on your bread and cheese can asexually produce spores that blow around in your kitchen and land on new food, starting the mold growth all over again. Some houseplants and strawberries produce little plantlets via asexual reproduction, while other plants reproduce asexually via bulbs, corms, and tubers. Even some animals, such as jellies and anemones, grow and then split asexually into two new individuals. In all of these organisms, asexual reproduction is a simple and efficient way of making offspring.

3. _____

2. _____

1. _____

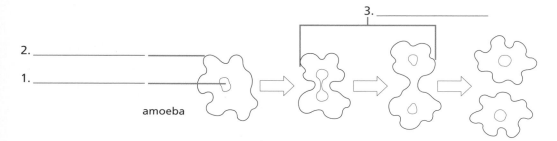

amoeba

Examples of Asexual Reproduction

4. _____

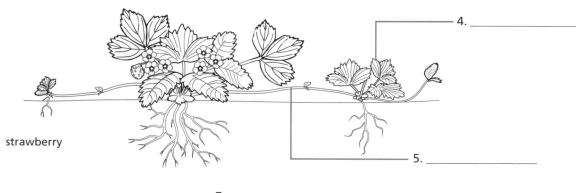

strawberry

5. _____

7. _____

6. _____

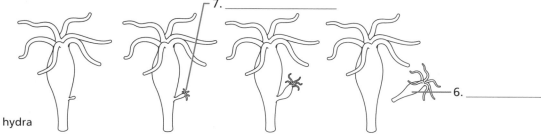

hydra

Answers

Chromosome Structure

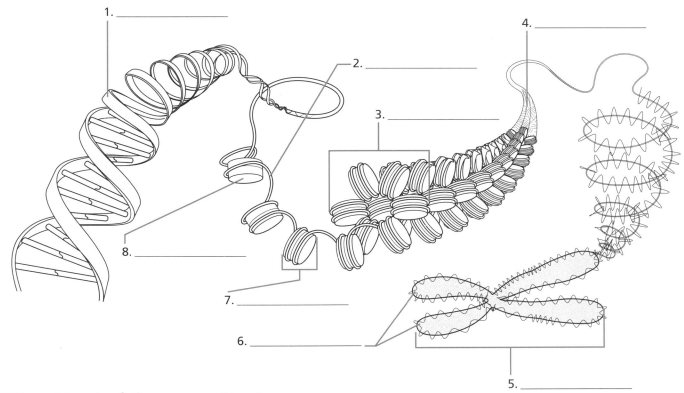

1. _____

2. _____

3. _____

4. _____

5. _____

6. _____

7. _____

8. _____

Different Levels of Chromosome Structure

Chromosomes are individual molecules of DNA wrapped around proteins. In eukaryotes, the negatively charged DNA wraps primarily around positively charged proteins called histones. A strand of DNA wraps twice around a clump of eight histone proteins and is held in place by a histone called H1. After a short stretch of linker DNA, the strand wraps again, forming a structure that's been described as "beads on a string." Each "bead" with its DNA wrap is called a nucleosome.

Additional packing of the chromosomes is necessary to fit all the DNA of a cell into the nucleus. Interactions between H1 proteins from different nucleosomes coil the chromosomes into fibers that are 30 nm wide. These fibers are then looped onto scaffolding proteins that organize the chromosome and hold it in place in the nucleus.

Even with these levels of packing, chromosomes are spread out through the nucleus and are barely visible as fine threads called chromatin during interphase. When cells enter M phase, the 30 nm fibers and scaffold proteins are twisted and folded further to condense the chromosomes tightly so that they can be sorted during cell division. This condensation makes chromosomes visible under a light microscope as individual structures consisting of pairs of sister chromatids.

Answers

1. DNA, 2. linker DNA, 3. solenoid, 4. chromatin, 5. condensed chromosome, 6. sister chromatids, 7. nucleosome, 8. histone

Mitosis

Eukaryotic cells divide as part of asexual reproduction, growth, and repair. To ensure that new cells get copies of all the chromosomes during cell division, cells use a controlled process of nuclear division called mitosis. Cells prepare for mitosis during interphase. When a cell gets a signal to divide, it leaves G1 and passes through S phase and G2 before beginning mitosis.

Mitosis is separated into five phases. During prophase, cells coil their chromosomes tightly (condensation). A system of microtubules called the mitotic spindle forms, and the nuclear membrane breaks down. During prometaphase, the chromosomes attach to the microtubules. Microtubules lengthen and shorten, tugging the chromosomes until they are all lined up in the middle of the cell at metaphase. During anaphase, identical sister chromatids release from each other, and the microtubules pull them to opposite poles of the cell. Telophase completes mitosis by reversing the events of prophase. After or during telophase, the cytoplasm of the cell divides by cytokinesis. In animal cells, a band of microfilaments contracts around the middle of the two cells, pinching them apart. In plants, microtubules deliver vesicles of membrane and wall materials to the middle of the cells to form a cell plate that eventually becomes the cell membrane and walls for the new cells.

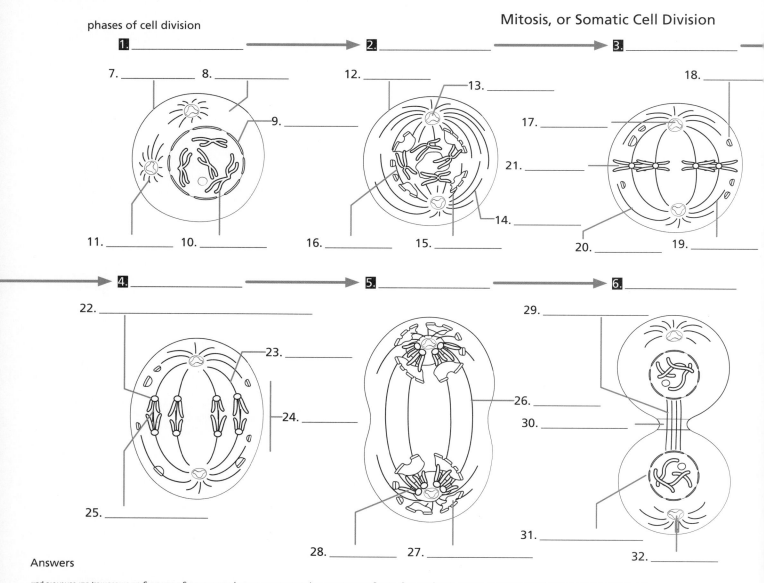

Answers

Cancer

Cancer is a genetic disease that results from the accumulation of mutations that affect cell division and behavior. Cancer begins in a single cell when a mutation affects a gene that regulates cell division, resulting in a cell that divides more frequently than others and begins to proliferate. Another similar mutation in a descendent of the same cell amplifies this effect, leading to the formation of a precancerous mass called a benign tumor or polyp. As the precancerous cells accumulate more mutations, their properties may change so that they no longer resemble their tissue of origin and they gain the ability to move through the body. These cells are malignant and have the potential to metastasize, or spread cancer throughout the body. Because cancer requires several mutations to occur in the same cell, the disease usually takes years to develop and is more common in older organisms.

Labels 1–4 refer to gene mutation types, and 5–9 refer to the type of cell growth.

Steps in the Transformation of a Normal Cell into a Cancer Cell

1. _____   2. _____   3. _____   4. _____

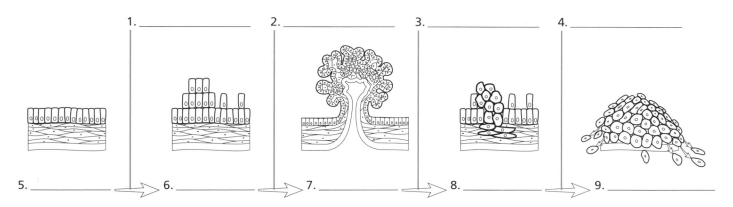

5. _____   6. _____   7. _____   8. _____   9. _____

The mutations associated with cancer occur in specific types of genes that normally regulate cell division. Proto-oncogenes, such as *ras*, produce proteins that stimulate cell division. Mutated *ras* proteins can overstimulate division. Tumor-suppressor genes, such as *APC* or *p53*, normally stop cell division so that other proteins can repair mistakes. When they don't function correctly, damaged cells continue to divide, leading to cells with changed properties. Normal genes associated with cell death produce proteins that either trigger apoptosis or prolong cell life by repairing the ends of chromosomes (telomeres). Mutations in these genes can grant immortality to a cancer cell.

Answers

1. *APC*, 2. *ras* and others, 3. *p53*, 4. others, 5. normal cells, 6. hyperproliferation, 7. benign polyp (adenoma), 8. malignant growth (carcinoma), 9. metastasis

Sexual Reproduction

Many types of organisms, including protists, fungi, plants, and animals, can reproduce sexually. During sexual reproduction, parents use a special type of cell division called meiosis to produce gametes, such as sperm and egg cells. Gametes are haploid (n), meaning they contain half of the genetic information of the parents. Two gametes fuse together in fertilization to produce a zygote, which is the first cell of a new organism. The zygote is diploid (2n), meaning it has the same number of chromosomes as the parents. If the organism is multicellular, the zygote divides by mitosis to produce the new cells of the organism. Cells of the new organism may specialize or differentiate to produce unique tissues during development.

Sexual reproduction is important because it allows for new combinations of traits in offspring. If a population is genetically identical, and environmental conditions change in a way that's harmful to that population, the chance that any individuals may survive could be low. With sexually reproducing organisms, however, each offspring represents a unique combination of genes, and thus traits, from the two parents. If adverse conditions confront a genetically diverse population, there's a chance that some individuals will have a combination of traits that enable them to survive and continue the species.

Major Steps in the Life Cycle of an Animal

For labels 1, 2, 4, and 7, include the number of chromosomes.

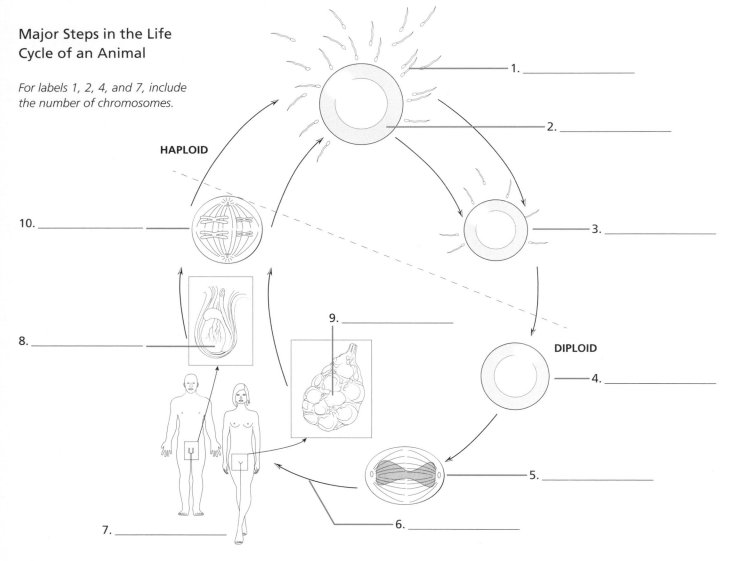

HAPLOID

DIPLOID

1. _____

2. _____

3. _____

4. _____

5. _____

6. _____

7. _____

8. _____

9. _____

10. _____

Answers

1. sperm (n = 23), 2. egg (n = 23), 3. fertilization, 4. zygote (2n = 46), 5. mitosis, 6. growth and development, 7. multicellular adults (2n = 46), 8. testis, 9. ovary, 10. meiosis

Overview of Meiosis

Meiosis is a special type of cell division that occurs in sexually reproducing species. The cells in which it occurs are diploid (2n), meaning they have two complete sets of chromosomes. The purpose of meiosis is to organize the chromosomes of a cell and separate them so that the resulting gametes have half the genetic information of the original cell. Meiosis proceeds carefully so that each gamete gets not just a random half the amount of DNA, but a complete set of chromosomes that contains one of each kind found in the parent cell. In other words, meiosis reduces the number of chromosomes from diploid (2n) to haploid (n).

Cells that receive a signal to enter meiosis follow the same cell cycle as cells that enter mitosis. From G1, they enter S phase and copy all their chromosomes. Each identical pair of sister chromatids remains attached. After they pass through G2, the cells enter meiosis. Meiosis uses two rounds of nuclear division, meiosis I and meiosis II. During meiosis I, the cells organize all the chromosomes, finding the matching pairs, or homologous chromosomes, and attaching them together to form tetrads. While the homologs stick together, they use crossing-over to swap pieces of DNA. The division in meiosis I separates the homologous pairs from each other, producing two haploid cells that still contain replicated chromosomes. The second round of division, in meiosis II, separates sister chromatids so that, in the end, meiosis produces four haploid gametes with unreplicated chromosomes.

Stages of Meiosis

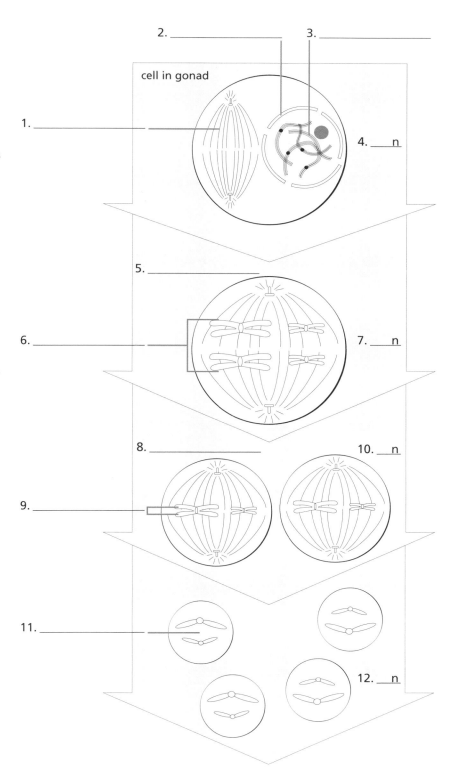

2. _____   3. _____

cell in gonad

1. _____

4. ___ n

5. _____

6. _____

7. ___ n

8. _____   10. ___ n

9. _____

11. _____

12. ___ n

Close-Up of Meiosis I

As in mitosis, scientists use certain key events to break meiosis I into stages. During prophase I, the cells get ready for nuclear division by condensing the chromosomes and forming the mitotic spindle. The nuclear membrane breaks down to allow the spindle to connect to the chromosomes via kinetochore proteins. The most distinctive event during prophase I is that the homologous chromosomes pair and stick to each other via attachment proteins. This event is essential because it allows the cell to organize all the pairs in preparation for carefully sorting them. While the homologs are paired, crossing-over occurs.

Once the homologs attach to the spindle, the microtubules lengthen and shorten to pull the chromosomes to the center of the cell. Scientists define metaphase I as the moment when the homologous pairs line up in the middle of the cell. During anaphase I, the homologous pairs let go of each other, and the spindle pulls one member of each pair to opposite poles of the cells. The nuclear membrane reforms during telophase I, the spindle breaks down, and the chromosomes uncoil. Depending on the species, cytokinesis may occur to divide the cell into two cells. Although these cells contain replicated chromosomes with sister chromatids, they're haploid because they contain only one of each homologous pair—in other words, one of each kind of chromosome in the cell.

Events of Meiosis I

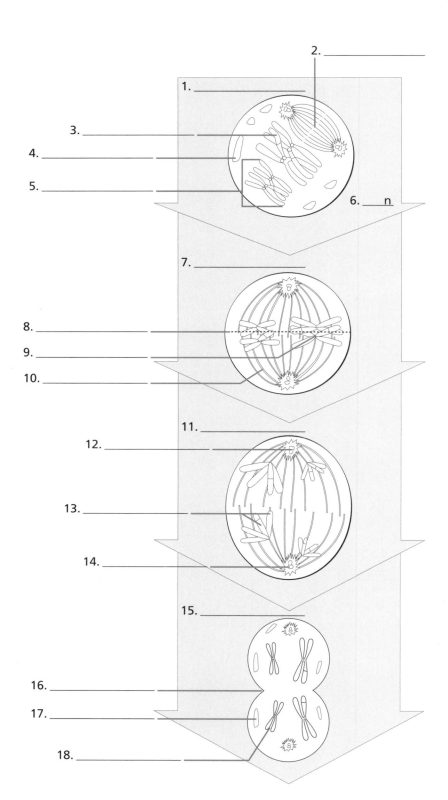

1. _____
2. _____
3. _____
4. _____
5. _____
6. __n__
7. _____
8. _____
9. _____
10. _____
11. _____
12. _____
13. _____
14. _____
15. _____
16. _____
17. _____
18. _____

Answers

1. prophase I, 2. mitotic spindle, 3. crossing-over, 4. nuclear membrane fragment, 5. homologous pair (tetrad), 6. 2, 7. metaphase I, 8. metaphase plate, 9. centromere, 10. microtubule, 11. anaphase I, 12. centriole, 13. sister chromatids, 14. centrosome (MTOC), 15. telophase I/cytokinesis, 16. cleavage furrow, 17. nuclear membrane fragment, 18. replicated chromosome

Close-Up of Meiosis II

Meiosis II begins with two haploid cells that have replicated chromosomes and then separates the sister chromatids to end with four haploid gametes that have unreplicated chromosomes. During prophase II, the cells prepare for nuclear division by condensing the chromosomes and forming the mitotic spindle. The nuclear membrane breaks down so that the microtubules can attach to the replicated chromosomes. The microtubules move the chromosomes so that they line up in the middle of the cell, marking metaphase II.

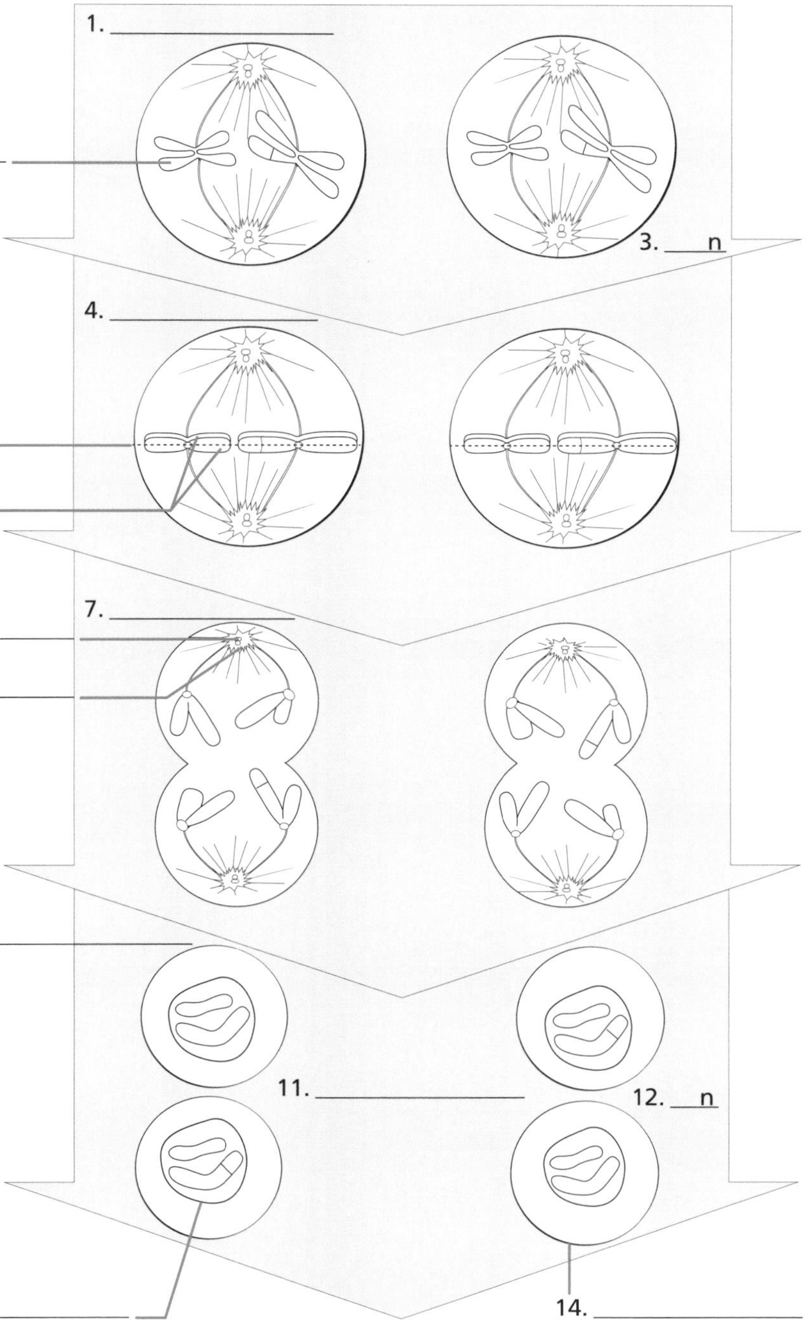

1. ____________
2. ____________
3. ___n
4. ____________
5. ____________
6. ____________
7. ____________
8. ____________
9. ____________
10. ____________
11. ____________
12. ___n
13. ____________
14. ____________

Events of Meiosis II

Sister chromatids separate from one another and the spindle pulls them to opposite sides of the cell during anaphase II. Telophase II reverses the events of prophase, reforming the nuclear membrane, uncoiling the chromosomes, and breaking down the spindle. Cytokinesis separates the cells, producing four haploid gametes.

Answers

1. prophase II, 2. replicated chromosome, 3. 1, 4. metaphase II, 5. metaphase plate, 6. sister chromatids, 7. anaphase II, 8. centriole, 9. centrosome (MTOC), 10. telophase II/cytokinesis, 11. gametes, 12. 1, 13. nuclear membrane, 14. plasma membrane

Close-Up of Crossing-Over

The most significant events during prophase I of meiosis are the pairing of homologous chromosomes and the crossing-over that occurs between them. Early in prophase I, the pairs of homologous chromosomes align and attach to each other by a network of proteins called the synaptonemal complex. Small breaks in the DNA can lead to points of genetic exchange between nonsister chromatids. These precise exchanges occur at multiple points along the chromosome, exchanging DNA from the same section of chromosome and creating chromatids that have a mixture of genes from both homologs. When the synaptonemal complex breaks down in late prophase I, the connection points between homologs are still visible as chiasma.

Events of Crossing-Over During Meiosis

1. _____

2. _____

3. _____

4. _____

5. _____

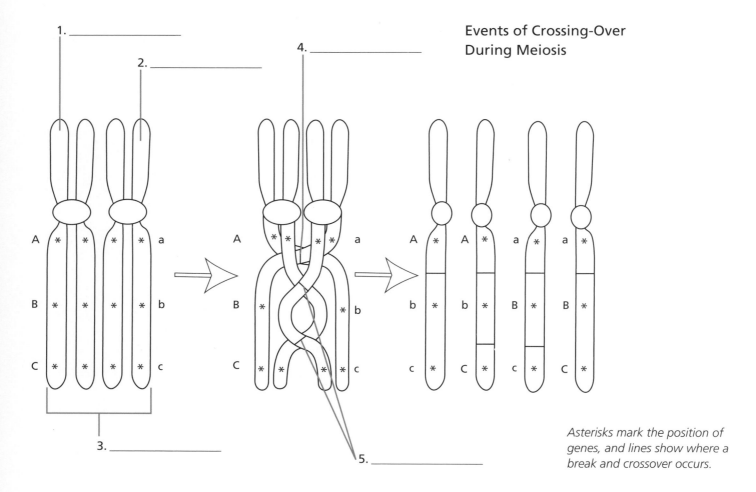

Asterisks mark the position of genes, and lines show where a break and crossover occurs.

The benefit of crossing-over is that it increases genetic variation in the offspring of sexually reproducing organisms. Not only do gametes receive unique combinations of chromosomes from the parent, but the chromosomes themselves may have unique mixtures of alleles from the parents' two homologs. Increasing variation in gametes leads to increased variation in offspring, which increases the chances that some offspring will survive to maturity.

Answers

1. maternal chromosome, 2. paternal chromosome, 3. homologous chromosomes (tetrad), 4. chiasma, 5. chiasma

Nondisjunction

Nondisjunction means a failure to separate. Specifically, it's the failure of chromosomes to separate during cell division. This is particularly important during meiosis, when nondisjunction can produce gametes that have the wrong number of chromosomes. If these gametes participate in fertilization, the result may be an aneuploid organism that has the wrong number of chromosomes in all of its cells. This organism may not be viable at all, or it could suffer a range of developmental abnormalities.

Nondisjunction can occur during either round of meiosis. When nondisjunction occurs during meiosis I, both members of a homologous pair go to the same pole during anaphase I. If the rest of meiosis happens normally, then the sister chromatids will separate during anaphase II. Ultimately, this produces two gametes with one extra chromosome and two gametes that lack a chromosome. When nondisjunction occurs during meiosis II, both sister chromatids of one chromosome go to the same pole during anaphase II. In this case, two of the resulting gametes will be normal, one will have an extra chromosome, and one will be missing a chromosome.

One of the most common outcomes of nondisjunction in humans is trisomy 21, which leads to Down syndrome. People with three copies of chromosome 21 may suffer from developmental delays and increased risk for certain diseases. The risk of producing an egg with the wrong number of chromosomes increases with the age of the mother, so older mothers face a higher risk of having a child affected by trisomy 21.

Outcome of Nondisjunction in Meiosis I and Meiosis II

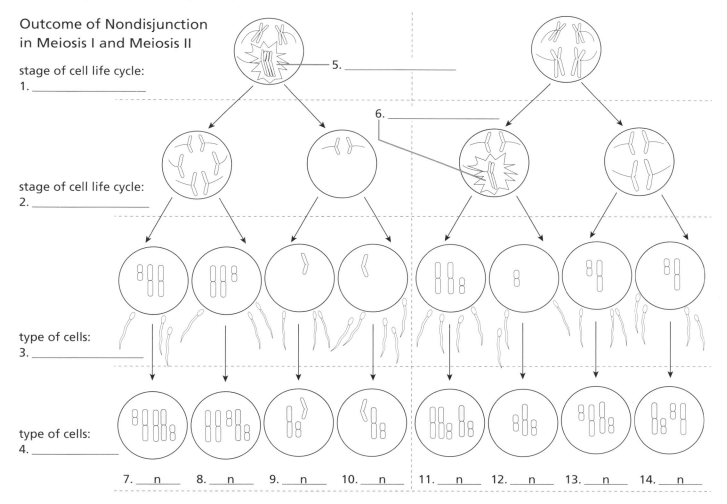

stage of cell life cycle:
1. _____

5. _____

6. _____

stage of cell life cycle:
2. _____

type of cells:
3. _____

type of cells:
4. _____

7. __n__   8. __n__   9. __n__   10. __n__   11. __n__   12. __n__   13. __n__   14. __n__

Answers

1. meiosis I, 2. meiosis II, 3. gametes, 4. zygotes, 5. nondisjunction, 6. nondisjunction, 7. 2 +1, 8. 2 +1, 9. 2 −1, 10. 2 −1, 11. 2 +1, 12. 2 +1, 13. 2, 14. 2

The Law of Segregation

People have probably always recognized that children inherit traits from their parents, but no one really understood how inheritance worked until the work of the Austrian monk Gregor Mendel in the mid-1800s. Before Mendel, the prevailing idea was that the traits of the parents blended together in their offspring.

Mendel's work was unique because, not only did he carefully conduct breeding experiments in peas, he also applied mathematics to predict and understand the patterns he was seeing. He started with true-breeding plants called parentals (P), such as a line that always produced purple flowers and one that always produced white flowers. He mated these to each other to produce a first generation called the F1, which all had purple flowers. Next, he bred members of the F1 to each other, creating a second generation called the F2. In the F2, he found both purple- and white-flowered plants in a ratio of three purple for every one white (3:1). Because the trait of white flowers returned in the F2, Mendel knew that the blended idea of inheritance couldn't be correct. Instead, he reasoned that there must be a distinct factor that controlled white flower color but that had been hidden in the F1 generation.

Mendel further tested his ideas and confirmed his Law of Segregation, which states that individuals have two copies of every gene, but they give only one allele for each gene to their gametes. Although meiosis and chromosomes hadn't been discovered when Mendel was carrying out his work, we now know that the segregation of alleles occurs during anaphase I of meiosis, when homologous pairs separate.

How Segregation of Alleles in Meiosis Affects the Genetics of Offspring

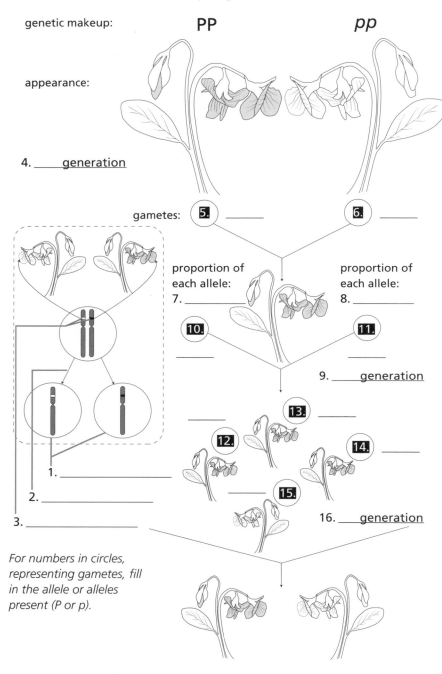

genetic makeup: **PP**   *pp*

appearance:

4. _____ generation

gametes: 5. _____   6. _____

proportion of each allele: 7. _____   proportion of each allele: 8. _____

10. _____   11. _____

9. _____ generation

13. _____

12. _____   14. _____

15. _____

16. _____ generation

1. _____

2. _____

3. _____

For numbers in circles, representing gametes, fill in the allele or alleles present (P or p).

17. phenotypic ratio: _____

The Law of Independent Assortment

When Gregor Mendel observed and tracked the inheritance of two traits at the same time, the patterns he observed revealed another aspect of how alleles separate during gamete formation. Mendel started with P plants that had pairs of opposite characteristics, such as plants that made smooth yellow seeds versus those that made wrinkled green seeds. When he crossed these plants to create the F1, all the F1 plants had the same characteristics as the smooth yellow parent. But when he crossed the F1 plants with each other, not only did he see wrinkled green peas, he also saw brand-new combinations of wrinkled yellow peas and smooth green peas. Because the parental combinations did not stay together in the offspring, Mendel concluded that the genes that control different traits must segregate independently from each other when parents make offspring. And, because he observed a 9:3:3:1 phenotypic ratio in the F2, he knew that this segregation was random.

With our modern understanding of chromosomes and meiosis, we now know that independent assortment happens during anaphase I of meiosis. In cells with multiple chromosomes, the separation of one pair of homologs doesn't influence the separation of another pair of homologs. This means that every time a parent makes gametes, their pairs of homologs may sort differently, giving different combinations of alleles to gametes and thus to offspring. In evolutionary terms, increased variation in offspring increases the chances that some offspring will survive to maturity, which is especially important in adverse environmental conditions.

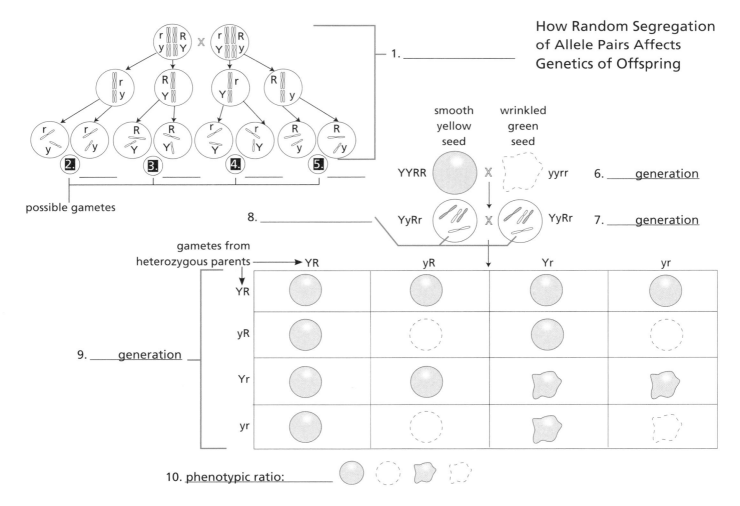

How Random Segregation of Allele Pairs Affects Genetics of Offspring

1. _____

possible gametes

2. _____ 3. _____ 4. _____ 5. _____

smooth yellow seed   wrinkled green seed

YYRR   X   yyrr   6. _____ generation

8. _____   YyRr   X   YyRr   7. _____ generation

gametes from heterozygous parents ⟶ YR   yR   Yr   yr

9. _____ generation

10. phenotypic ratio: _____

Answers

DNA Replication

DNA replication is the process cells use to copy their DNA prior to cell division. In eukaryotes, this occurs during S phase of the cell cycle (see p. 53). Cells copy their chromosomes by separating the two strands of each DNA double helix and using each as the template for the synthesis of a new strand. The enzyme DNA polymerase III uses base-pairing rules to match nucleotides for the new strands with the nucleotides in the parent strand. When DNA replication is complete, each chromosome has been doubled. In eukaryotes, the two copies of each chromosome remain attached to each other as sister chromatids until the cells reach M phase of the cell cycle. Because each new chromosome consists of half original DNA and half new DNA, scientists say that DNA replication is semiconservative.

The areas where DNA is actively being copied are visible in cells as openings called replication bubbles, which have a replication fork on either side. As DNA polymerase III builds the new strands, it places each nucleotide into the chain by catalyzing covalent bond formation between the 3' end of an existing nucleotide and the 5' end of the incoming nucleotide. For one new strand, called the leading strand, DNA polymerase adds nucleotides continuously. For the other new strand, called the lagging strand, the enzyme must add nucleotides in short segments called Okazaki fragments.

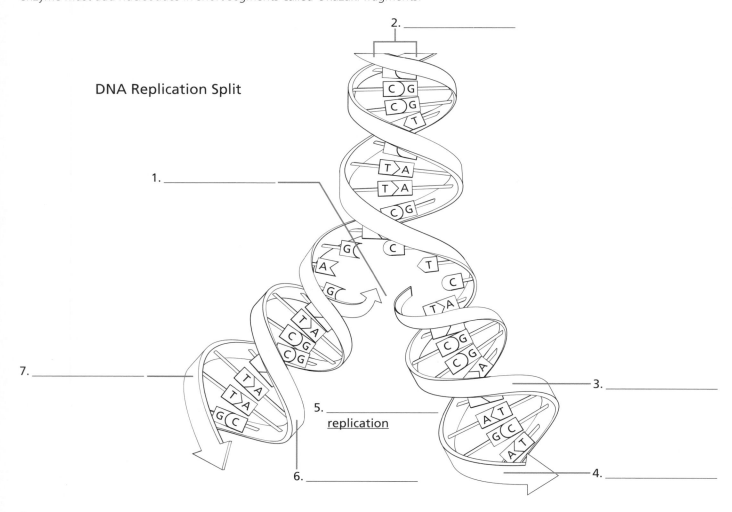

DNA Replication Split

2. _____

1. _____

7. _____

5. _____
replication

3. _____

6. _____

4. _____

Answers

Close-Up of a Replication Fork

DNA polymerase III is part of a team of enzymes and other proteins that work together as a large enzyme complex to replicate DNA. Two of these enzyme complexes begin at an origin of replication and move in opposite directions along the parental DNA. As the complexes move, each protein does its part in an ordered sequence of events. A snapshot in time of a replication fork illustrates the role of each protein.

First, helicase separates the parental DNA by breaking the hydrogen bonds between the base pairs. Next, single-strand binding proteins attach to the parental strands to stop them from reattaching to each other. Primase uses base-pairing rules to form short pieces of complementary RNA, called primers, on the parental strands. These primers provide a starting place for DNA polymerase III to begin synthesizing the new strands of DNA. Once DNA polymerase III has the 3′ end of a primer to build on, it begins to use base-pairing rules to make a new partner strand for each parental strand. Another enzyme, DNA polymerase I, replaces the RNA nucleotides in the primer sections with DNA nucleotides. Finally, the enzyme ligase catalyzes bond formation in any places where two fragments in the new strands come together.

DNA Replication Fork

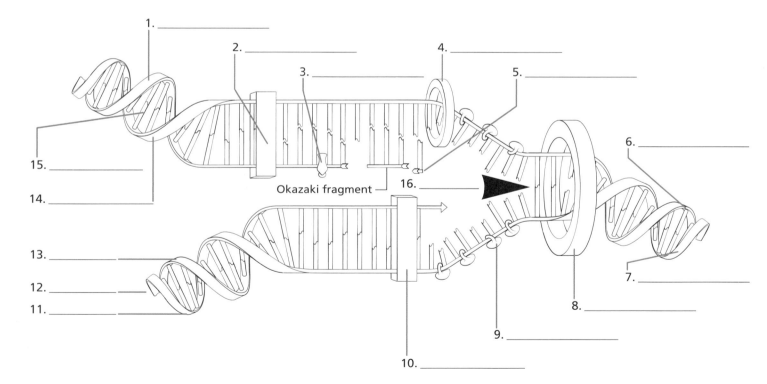

1. _____
2. _____
3. _____
4. _____
5. _____
6. _____
7. _____
8. _____
9. _____
10. _____
11. _____
12. _____
13. _____
14. _____
15. _____
16. _____

Okazaki fragment

Transcription is the process cells use to copy the information in DNA into RNA molecules. Cells produce several different types of RNA molecules that perform different functions. Three of these, rRNA, tRNA, and mRNA, are necessary for cells to build proteins using the process of translation. Enzymes called RNA polymerases use base-pairing rules to build these RNA molecules based on their genes located in the DNA.

    Special DNA sequences called promoters mark the locations of genes within chromosomes. When a cell gets a signal that it needs a certain RNA molecule, regulatory proteins help RNA polymerase locate and bind to specific locations in the correct promoters. This binding orients RNA polymerase as to which DNA strand to use as a template and which direction to travel along the DNA.

    RNA polymerase unwinds a section of the DNA and makes an RNA molecule complementary to the template strand. The sequence of this RNA molecule will be almost identical to that of the DNA strand that's complementary to the template strand, the only difference being the presence of uracil (U) in the RNA versus thymine (T) in the DNA. Because it shows the code that will end up in the RNA transcript, the complementary DNA strand is called the coding strand. RNA polymerase moves along the DNA, transcribing RNA until it reaches a special sequence called the transcription terminator. Depending on the cell type, various mechanisms cause RNA polymerase to end transcription.

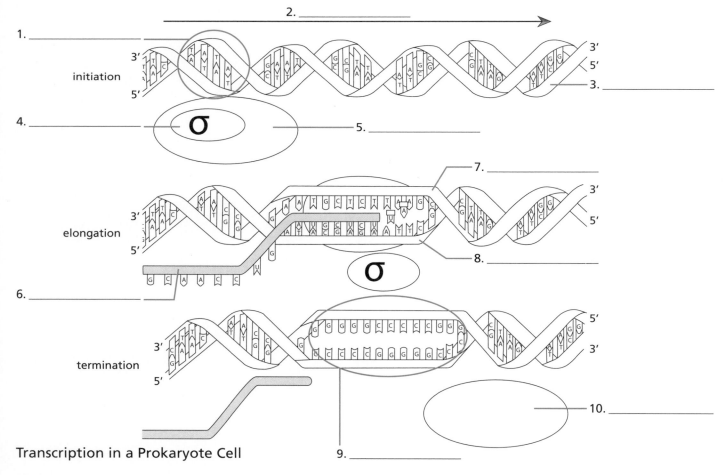

Transcription in a Prokaryote Cell

Answers

RNA Processing

In eukaryotes, the products of transcription are primary transcripts that need modification before they are functional RNA molecules. RNA processing refers to the finishing modifications that eukaryotic cells use to convert primary transcripts to functional RNA.

RNA splicing removes sections of RNA that aren't needed in the final RNA. Eukaryotic genes contain alternating sections of DNA called introns and exons. Only the code in the exons is actually used, or expressed, in the final RNA. During transcription, particles called small nuclear ribonucleoproteins (snRNPs, or "snurps") attach to the primary transcript at the boundaries between exons and introns. The snRNPs join together, forming a spliceosome that cuts the RNA at the exon-intron boundaries, forms the introns into lariats (stem plus loop), releases the introns, and joins the remaining exons together. Finished RNA molecules are thus much shorter than the primary transcripts.

If the primary transcript is destined to be a molecule of mRNA, two additional finishing touches are needed. While the pre-mRNA is still being transcribed, cells attach a modified guanine nucleotide to its 5' end. At the 3' end, cells attach a poly(A) tail made of 100–250 adenine nucleotides. The 5' cap and poly(A) tail protect the RNA from enzymatic breakdown and are necessary for the initiation of translation.

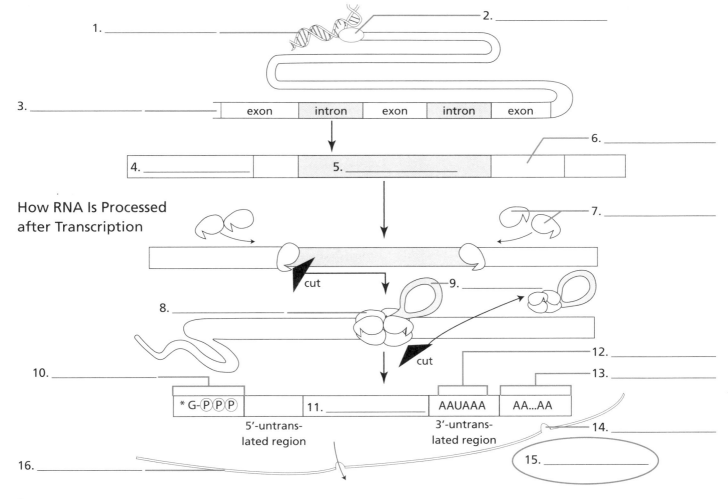

1. _____

2. _____

3. _____

exon | intron | exon | intron | exon

4. _____

5. _____

6. _____

How RNA Is Processed after Transcription

cut

7. _____

8. _____

9. _____

cut

10. _____

* G-(P)(P)(P)

11. _____

5'-untranslated region

AAUAAA | AA...AA

3'-untranslated region

12. _____

13. _____

14. _____

15. _____

16. _____

Answers

Translation

Cells use translation to build proteins according to the code carried by mRNA molecules. Translation occurs at ribosomes and requires the action of tRNA and protein factors. Ribosomes provide structure that organizes the process and also catalyze bond formation between amino acids in the growing polypeptide chains. tRNA molecules carry amino acids into the ribosome, placing each one in the correct location based on hydrogen binding between bases in the tRNA and mRNA. Each tRNA molecule has a special section of three nucleotides called the anticodon, which pairs with groups of three nucleotides called codons in the mRNA.

　　Translation begins when the small subunit of a ribosome binds to the mRNA at the ribosomal binding sequence. Next, the initiator tRNA pairs with the start codon, bringing the first amino acid into place. The large subunit binds to complete the ribosome, creating three internal pockets called the E, P, and A sites.

　　Cells repeat a cycle that brings amino acids into the ribosome and adds them one at a time to the elongating polypeptide chain. tRNAs bring their amino acids into the A site. The ribosome catalyzes peptide bond formation, attaching the growing polypeptide chain onto the newest amino acid. Proteins shift the ribosome and mRNA relative to one another, moving the growing chain to the P site, opening up the A site, and pushing a tRNA out through the E site. The cycle continues until a stop codon is in the A site, which signals a protein called release factor to enter the A site and release the polypeptide chain.

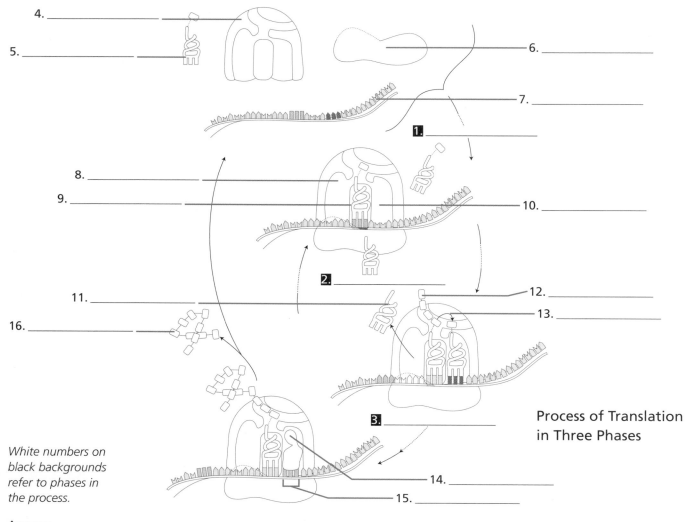

4. _____ _____

5. _____ _____

6. _____

7. _____

1. _____

8. _____ _____

9. _____

10. _____

2. _____

11. _____

16. _____

12. _____

13. _____

3. _____

Process of Translation in Three Phases

White numbers on black backgrounds refer to phases in the process.

14. _____

15. _____

Answers

Mutation

Mutations are changes in the DNA code of an organism. If the change affects just one or a few nucleotides, scientists call this a point mutation. The impact of point mutations in protein-encoding genes varies from no effect to severe, depending on the type and location of the mutation.

Silent mutations are base substitutions that have no effect. This is possible because the genetic code is redundant, meaning that multiple codons may represent the same amino acid. If one base is substituted for another, but the result is an alternative codon for the same amino acid, then the protein doesn't change at all.

Missense mutations occur when base substitutions do change amino acids. The severity of the impact depends on the properties of the new amino acid compared to the original, and how much impact the change has on protein structure and function.

Nonsense mutations result from base substitutions that create stop codons. These mutations lead to early termination of translation and usually have severe impacts on protein function.

In addition to base substitutions, cells sometimes accidentally insert or delete bases from the DNA. These types of mutations are typically severe because they alter the reading frame of the mRNA, shifting the original groupings of nucleotides into codons. After the point of insertion or deletion, the protein may be completely different from the original.

Types of Point Mutations

Label the amino acids using the following abbreviations: arginine (Arg), glutamic acid (Glu), glycine (Gly), lysine (Lys), phenylalanine (Phe), threonine (Thr), tyrosine (Tyr).

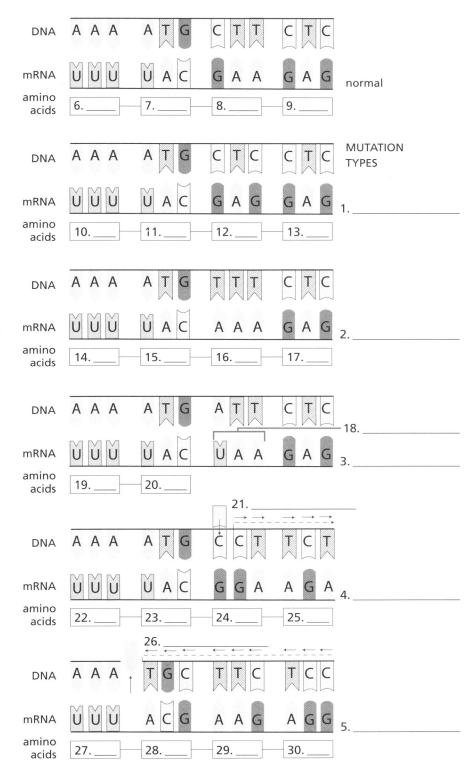

Answers

1. silent, 2. missense, 3. nonsense, 4. insertion, 5. deletion, 6. Phe, 7. Tyr, 8. Glu, 9. Glu, 10. Phe, 11. Tyr, 12. Glu, 13. Glu, 14. Phe, 15. Tyr, 16. Lys, 17. Glu, 18. stop codon, 19. Phe, 20. Tyr, 21. frameshift, 22. Phe, 23. Tyr, 24. Gly, 25. Arg, 26. frameshift, 27. Phe, 28. Thr, 29. Lys, 30. Arg

Control of Gene Expression in Prokaryotes

Scientists first figured out how cells control gene expression by studying the *lac* operon in the bacterium *Escherichia coli*. An operon is a section of DNA that consists of multiple genes that share a common promoter. The *lac* operon contains the blueprints for proteins that *E. coli* needs to break down the sugar lactose. *Escherichia coli* builds these proteins only when lactose is available and other food sources aren't. Scientists figured out that proteins bind to regulatory regions of the DNA in order to turn transcription of these genes on and off in response to environmental conditions.

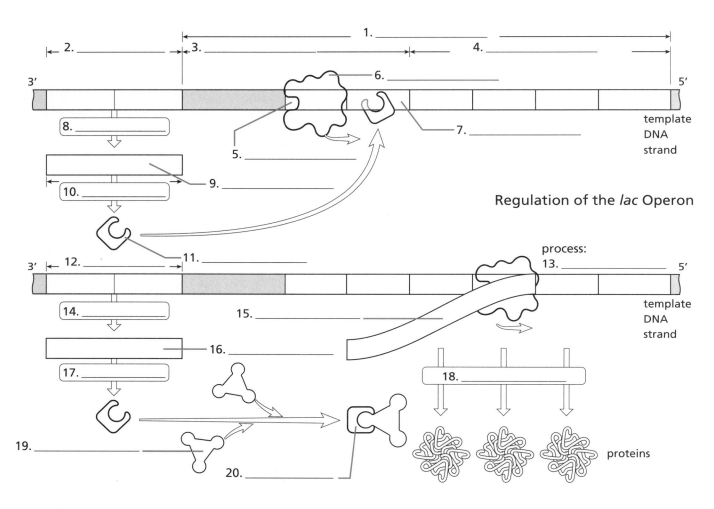

Regulation of the *lac* Operon

One of the key regulatory proteins for this operon is the *lac* repressor protein. This protein has a DNA-binding site that binds to the operator, a stretch of DNA between the *lac* promoter and the genes in the operon. When lactose isn't available, the *lac* repressor is active and binds to the operator, physically blocking RNA polymerase from being able to bind the promoter and begin transcription. When lactose becomes available, it binds to an allosteric site on the repressor protein, changing its shape so that it can no longer bind to the operator. This clears the way for RNA polymerase to bind to the promoter and transcribe the genes so that *Escherichia coli* can break down lactose. Once the lactose is gone, the repressor will again bind to the DNA and stop transcription. Because this operon is turned off unless lactose turns it on, scientists say that the *lac* operon is an inducible operon and that lactose is the inducer.

Answers

Control of Gene Expression in Eukaryotes

Eukaryotes have many possible control points along the journey from a gene in the nucleus to the expression of that gene's product in the cell. First, in order for transcription to occur, the genes must be accessible to RNA polymerase. This may require chromatin remodeling to loosen connections between DNA and histones. Then, just as in prokaryotes, proteins called transcription factors bind to regulatory sequences of DNA in order to control access of RNA polymerase to the genes. Activators bind to enhancer sequences of DNA in order to stimulate transcription, whereas repressors bind to silencer sequences to slow transcription.

Even if a cell transcribes a gene, it still has several posttranscriptional opportunities to control the expression of the gene product. Cells can alter the splicing of a primary transcript in order to produce many different variations of finished RNA molecules. This alternative splicing enables cells to make many variations of a gene product from a single gene.

Cells can also regulate gene expression by translational control, such as using RNA interference. Tiny RNA molecules attach to protein complexes and then bind to complementary sequences on mRNA molecules, blocking their translation or even triggering their destruction. Translational control can also occur by regulation of any of the many proteins that participate in the process.

Finally, once a protein is present in a cell, it can still be controlled by posttranslational methods. Cells can add carbohydrates or phosphate groups to proteins, altering their activity. Proteins that aren't needed any longer can also be targeted for destruction.

Mechanisms for Gene Regulation in Eukaryotes

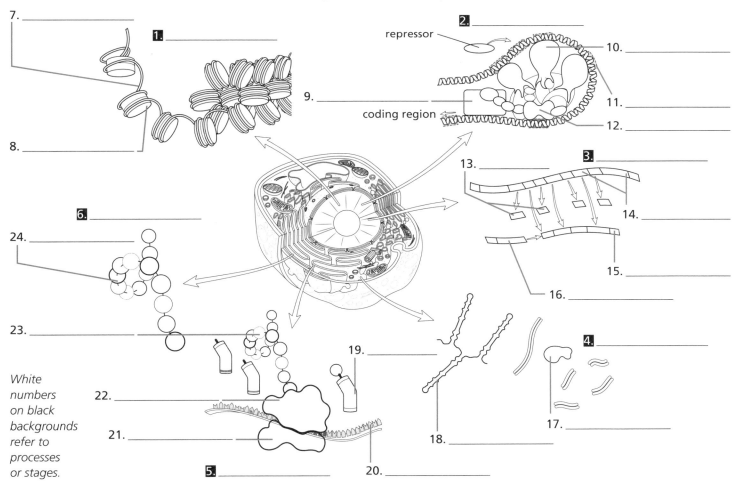

7. _____

1. _____

2. _____

repressor

10. _____

9. _____

coding region

11. _____

12. _____

8. _____

13. _____

3.

6. _____

14. _____

24. _____

15. _____

16. _____

23. _____

4. _____

19. _____

17. _____

White numbers on black backgrounds refer to processes or stages.

22. _____

21. _____

18. _____

5. _____

20. _____

Answers

Restriction Enzymes

Restriction enzymes, or restriction endonucleases, are proteins that cut double-stranded DNA at specific sequences called restriction sites. Each enzyme recognizes and cuts at a difference sequence. Some restriction enzymes make a blunt cut that severs both DNA strands at the same place, but others make a staggered cut that leaves a bit of single-stranded DNA extending from the cut site. The single-stranded portions of the two original strands are complementary, so they tend to reattach to each other via hydrogen bonds. Because of this, scientists call them "sticky ends."

In nature, bacteria produce restriction enzymes to aid them in defense against viruses. When a bacteriophage injects its DNA into a bacterial cell, the bacterium may respond by producing restriction enzymes that break up the viral DNA and prevent the virus from reproducing. When scientists discovered these enzymes in bacteria, they realized that the enzymes could be valuable tools in the laboratory. Now, scientists use restriction enzymes to break up DNA molecules as part of techniques such as genetic engineering. When scientists want to combine DNA from two sources, they cut both types of DNA with the same restriction enzyme in order to generate matching sticky ends. This allows the two types of DNA to stick together, after which they can be joined permanently using DNA ligase.

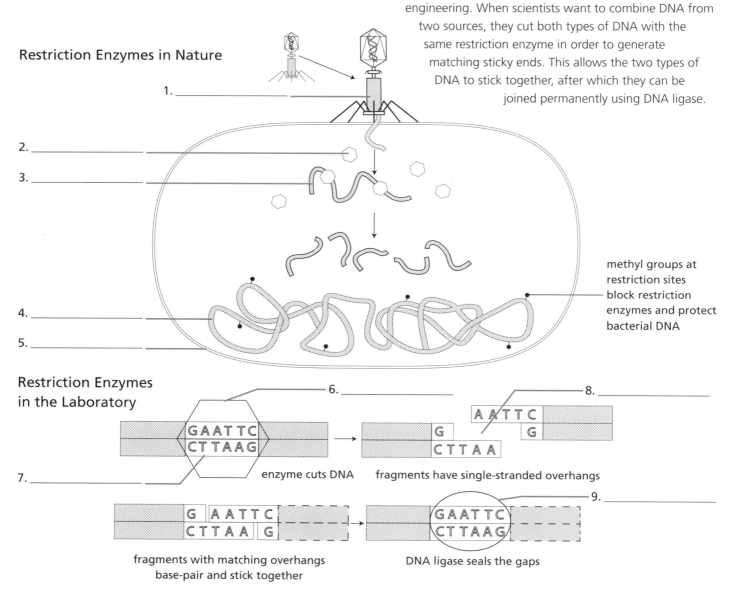

Restriction Enzymes in Nature

1. _____

2. _____

3. _____

methyl groups at restriction sites block restriction enzymes and protect bacterial DNA

4. _____

5. _____

Restriction Enzymes in the Laboratory

6. _____

8. _____

GAATTC
CTTAAG

G
CTTAA

AATTC
G

7. _____

enzyme cuts DNA

fragments have single-stranded overhangs

9. _____

G | AATTC
CTTAA | G

GAATTC
CTTAAG

fragments with matching overhangs base-pair and stick together

DNA ligase seals the gaps

Answers

Gel Electrophoresis

Gel electrophoresis uses an electrical current to separate molecules based on their size. Scientists load proteins or DNA molecules into one end of a gel, and then run a current through the gel. The current attracts the charged groups on the molecules, pulling the molecules through the gel. Smaller molecules move more easily and travel greater distances than larger molecules in a set amount of time. Scientists compare the position of the bands in an unknown sample to the position of known molecules called markers in order to determine the sizes of unknown molecules. Gel electrophoresis plays an important role in many techniques, including DNA fingerprinting and DNA sequencing.

Several steps are necessary for gel electrophoresis. First, scientists pour a liquefied gelatinous molecule, such as agarose, into a gel mold (A and B). They insert a comb into one end of the gel to create pockets called wells (B and C). After the gel solidifies, they place it into an electrophoresis box that contains running buffer. Next, scientists combine the samples with loading buffer and load them into the wells along with markers for comparison (D).

Once the samples are loaded (E), scientists turn on the electrical current, causing the samples to move through the gel in lanes that grow from each well. DNA is negatively charged, so it moves toward the positive electrode. When scientists shut off the current, the molecules all stop in place (F), forming collections of molecules of the same size called bands. Finally, scientists use stains to reveal the location of the bands in the gel. Ultraviolet light may be needed to activate the stains.

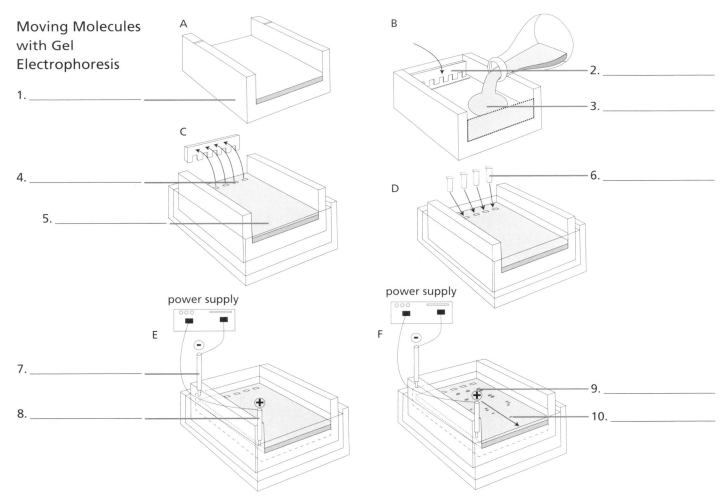

Moving Molecules with Gel Electrophoresis

A

1. _____

B

2. _____

3. _____

C

4. _____

5. _____

D

6. _____

power supply

E

7. _____

8. _____

power supply

F

9. _____

10. _____

Answers

1. casting tray, 2. comb, 3. agarose solution, 4. wells, 5. gel, 6. DNA samples, 7. negative electrode (cathode), 8. positive electrode (anode), 9. longer molecules, 10. shorter molecules

Polymerase Chain Reaction

The polymerase chain reaction (PCR) is one of the most valuable tools in biology today because it allows scientists to make billions of copies of a single DNA molecule in just a few hours. The basic mechanism behind PCR is the exponential replication of DNA molecules: one molecule becomes two, which become four, which become eight, and so on. The process requires the ingredients for DNA replication: DNA polymerase, template DNA, nucleotides (deoxyribonucleotide triphosphates, NTPs), and primers. Scientists place these materials along with buffer into PCR tubes and then place the tubes in a thermocycler. The thermocycler uses cycles of heating and cooling to separate DNA strands and then optimize DNA replication. Because the thermocycler heats the samples to high temperatures to separate the DNA, scientists use a special heat-stable DNA polymerase called *Taq* polymerase.

The PCR cycle has three steps. First, the thermocycler heats up in order to denature, or separate, the template strands of DNA. Next, the thermocycler cools so that the primers can anneal, or form hydrogen bonds, with the template DNA. Finally, the thermocycler raises the temperature to optimal conditions for *Taq* polymerase so the enzyme can produce new partners for the original strands. By repeating this cycle about 30 times, the thermocycler produces enough DNA for many different processes, including DNA fingerprinting, DNA sequencing, genetic testing, and genetic engineering.

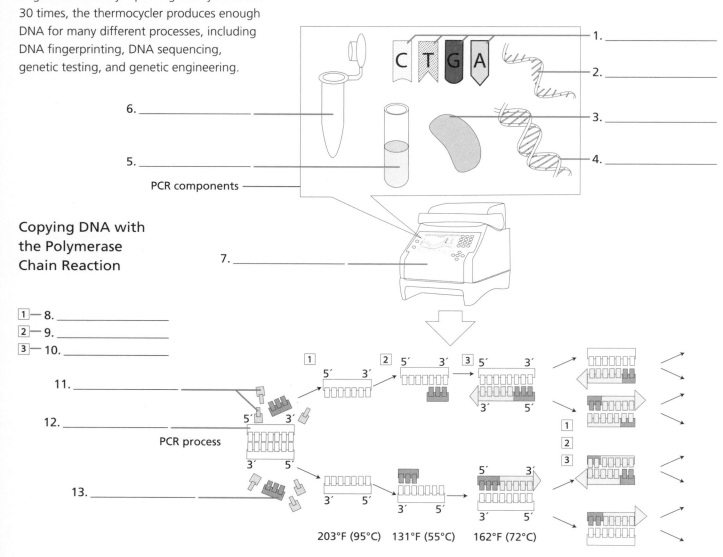

Copying DNA with the Polymerase Chain Reaction

1. _____
2. _____
3. _____
4. _____

6. _____
5. _____

PCR components

7. _____

8. _____
9. _____
10. _____

11. _____
12. _____

PCR process

13. _____

203°F (95°C)   131°F (55°C)   162°F (72°C)

DNA Fingerprinting

DNA fingerprinting creates a profile of an organism's DNA that is represented by a pattern of bands in a gel. First, scientists extract DNA from the organism to be profiled. Next, they typically use PCR to amplify several specific regions of the DNA that are polymorphic (variable) in that type of organism. Humans, for example, have areas in their DNA where short sequences, called short tandem repeats (STRs), repeat themselves. The number of times one of these STRs repeats itself varies from person to person. To create a DNA fingerprint, scientists use PCR to amplify several of these STR areas at once, generating a sample of DNA that contains fragments of different lengths. Scientists then use gel electrophoresis to separate the fragments based on their size.

The pattern of bands that results from gel electrophoresis is the DNA fingerprint. The pattern from two different organisms can be compared, for example, to resolve questions of paternity or to compare a DNA sample from a crime scene to a suspect. The chance that two organisms would match for multiple polymorphisms at once is the product of all of the independent probabilities. So, while a person might have a one in 10,000 chance of randomly matching another person for one polymorphism, the probability that they would match for multiple sites can shrink to as little as one in a quintillion (10^{18}).

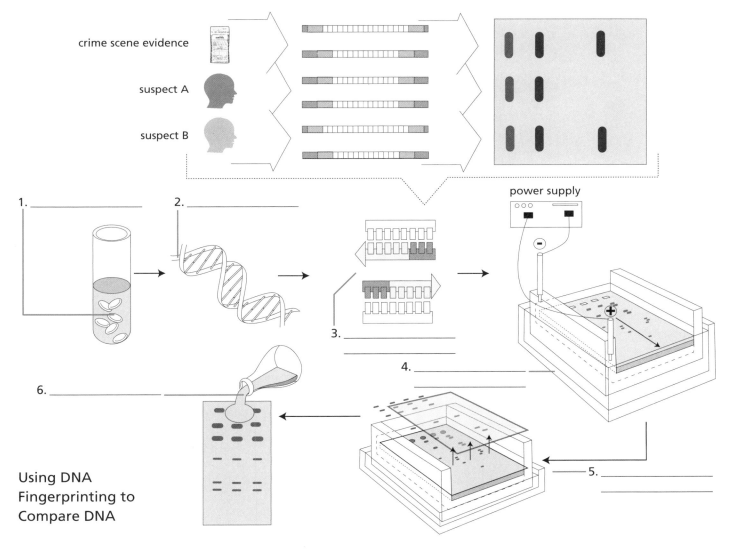

crime scene evidence

suspect A

suspect B

power supply

1. _____

2. _____

3. _____
   _____

4. _____
   _____

5. _____
   _____

6. _____

Using DNA Fingerprinting to Compare DNA

Answers

DNA Sequencing

DNA sequencing determines the sequence, or order, of the bases in a DNA molecule. Modern DNA sequencing uses a modification of PCR to create many partial copies of a target DNA molecule. The final nucleotide in each partial copy is labeled so that scientists can identify the complete sequence by sorting all the partial copies by size.

To produce partial copies of the target DNA sequence, scientists add special types of nucleotides, called dideoxynucleotide triphosphates (ddNTPs), to the normal ingredients of a PCR reaction mix. Once DNA polymerase incorporates a ddNTP into a growing molecule, replication stops. Scientists make four different PCR reaction mixes, each one with a different ddNTP (ddATP, ddGTP, ddTTP, ddCTP). In each tube, PCR will produce billions of partial copies of the DNA sequence that were randomly stopped at all the positions that contain each base. Gel electrophoresis sorts all of the partial copies by size. Scientists determine the order of the bases in the target gene by "reading" the bands from smallest to largest. In DNA sequencing machines, electrophoresis occurs in a capillary tube with an attached laser that activates the fluorescent dyes on the ddNTP molecules. As the DNA fragments flow through the tube, a computer reads the fluorescent signals to identify the final base in each fragment and construct the sequence.

Reading the Code with DNA Sequencing

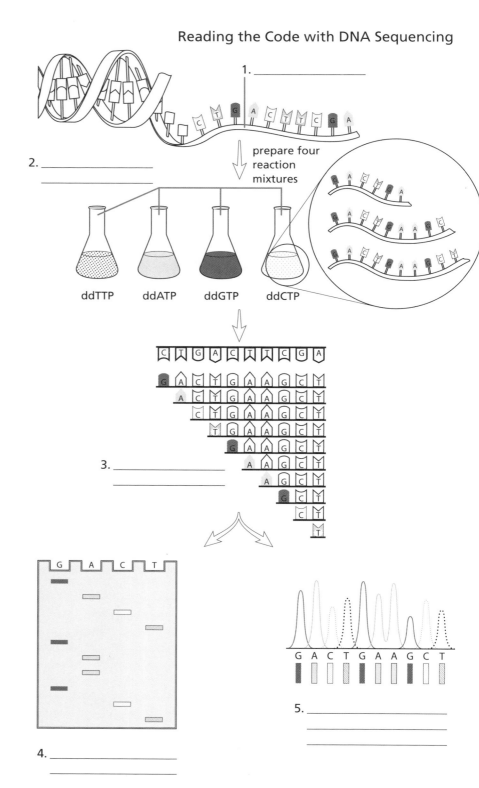

1. _____

2. _____

prepare four reaction mixtures

ddTTP   ddATP   ddGTP   ddCTP

3. _____

4. _____

5. _____

Answers

1. DNA to be sequenced, 2. add dideoxynucleotide triphosphates (ddNTPs), 3. PCR using nucleotides and ddNTPs, 4. gel electrophoresis separates DNA molecules, 5. computer reads fluorescent tags on the bands to construct sequence

Cloning a Gene

Cloning is the term used to describe making copies of something, such as an animal, a cell, or a gene. To clone a gene, genomic DNA is extracted from the organism of interest and then cut into fragments. A DNA molecule called a vector is used to put the fragments into cells of something that grows easily in the lab, such as bacteria or yeast. Vectors typically include antibiotic resistance genes, such as the *ampR* gene, and reporter genes that help scientists identify which cells successfully picked up genomic DNA. The cells containing the fragments are a genomic library, a living repository of the DNA from the organism of interest. Scientists use various methods to screen the cells in the library to find which ones carry the gene of interest. Once they identify the right cells, they can grow these to make many copies of the gene.

To make the genomic library, scientists cut the genomic DNA and the vector DNA with the same restriction enzyme to create complementary sticky ends. The restriction site in the vector occurs in the middle of a reporter gene, a gene for an enzyme that turns its substrate into a colored product, such as the *lacZ* gene. When the genomic fragments mix with the vector, some vectors will pick up a fragment, disrupting the gene for the enzyme. Next, scientists insert the vectors into the cells and then grow the cells with an antibiotic so that only cells that picked up the vector can grow. Scientists can also identify which cells contain genomic DNA, because these cells won't produce a color-changing reaction.

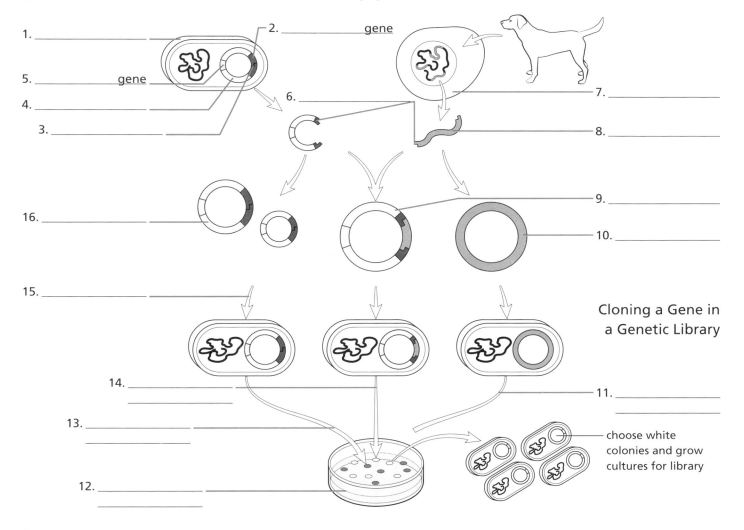

1. _____
2. _____ gene
5. _____ gene
4. _____
3. _____
6. _____
7. _____
8. _____
9. _____
10. _____
16. _____
15. _____
14. _____
13. _____
12. _____
11. _____

choose white colonies and grow cultures for library

Cloning a Gene in a Genetic Library

Answers

1. bacterial cell, 2. *lacZ*, 3. restriction site, 4. vector (plasmid), 5. *ampR*, 6. sticky ends, 7. cell from dog, 8. dog DNA fragments, 9. recombinant plasmid, 10. dog DNA ligates with itself, 11. cells die; no antibiotic resistance, 12. media with antibiotic and substrate for *lacZ*, 13. colonies are gray; *lacZ* gene intact, 14. colonies are white; *lacZ* gene interrupted, 15. media with antibiotic, 16. non-recombinant plasmid

Clustered Regularly Interspaced Short Palindromic Repeats

Clustered regularly interspaced short palindromic repeats (CRISPR, pronounced "crisper") refers to a DNA targeting system that scientists can use to locate and edit genes. CRISPR is a region of DNA that contains short palindromic sequences separated by spacer sequences. The spacer DNA is the key to the targeting system: when cells transcribe this DNA into RNA, the RNA locates and binds to its complementary sequence in DNA. The CRISPR system also contains genes for *Cas* enzymes, which cut the DNA at the targeted location. Scientists can use the CRISPR system to disrupt, activate, and even edit genes, which may lead to new therapies for genetic diseases. Scientists have already created thousands of RNA sequences, called guide RNAs (gRNAs), that provide targeting to genes of interest.

In nature, prokaryotes use the CRISPR system to defend themselves against viruses. When a prokaryote is attacked by a virus, the cell adds pieces of the viral DNA as new spacers in its CRISPR region. If the same virus attacks again, transcription of CRISPR will produce RNA molecules complementary to the viral DNA from the spacers. This RNA will lead the *Cas* enzymes to the viral DNA so they can cut it and deactivate the virus.

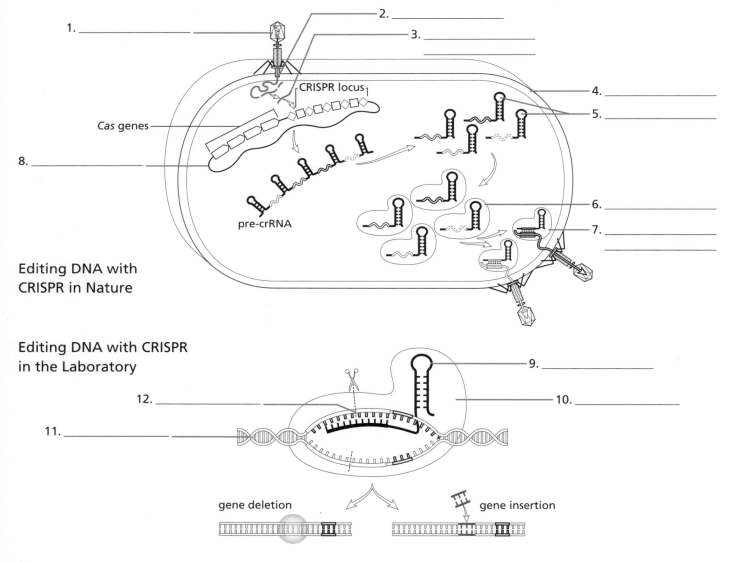

1. _____
2. _____
3. _____
   _____
4. _____
5. _____
6. _____
7. _____
8. _____

CRISPR locus

Cas genes

pre-crRNA

Editing DNA with CRISPR in Nature

Editing DNA with CRISPR in the Laboratory

9. _____
10. _____
11. _____
12. _____

gene deletion

gene insertion

Answers

Gene Therapy

Gene therapy refers to the process of introducing a copy of a gene into cells as a way to treat genetic disease. Genetic diseases result from mutations in genes that lead to improper protein function. Although some genetic diseases can be treated, the only way to truly cure a genetic disease would be to replace the faulty gene. Scientists are working to create safe vectors like modified viruses that can deliver genes to cells without causing harm to the patient. Another challenge lies in figuring out how to get the gene to enough of the affected cells to make a difference. Although many barriers remain, scientists have begun to make progress in this type of therapy.

Two methods have had some success in introducing genes into defective cells. One method is to remove cells from the patient, use a vector such as a modified virus to introduce a working copy of the gene, and then return the cells to the patient's body. Scientists have used this method to repair white blood cells in order to restore the immune systems of people with severe immune deficiency. The second method is to introduce a vector carrying the therapeutic gene directly into the patient's body. With this method, scientists have been able to partially correct some forms of inherited blindness and attack cancer cells.

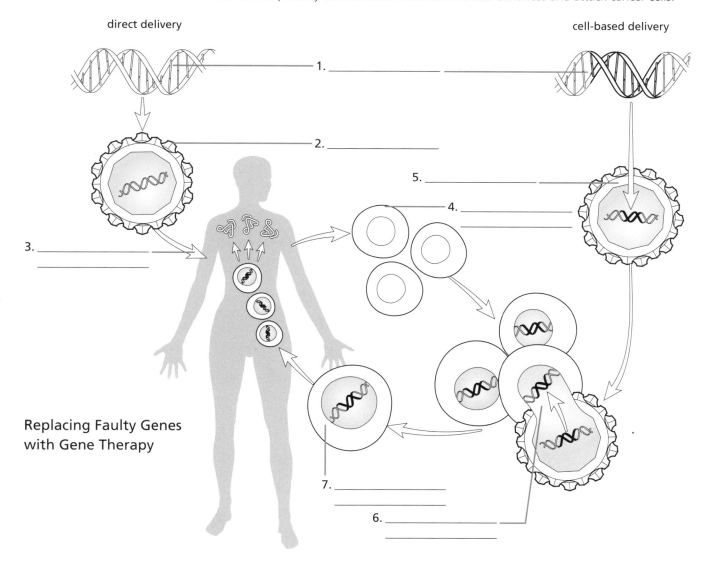

direct delivery

cell-based delivery

1. _____ _____

2. _____

5. _____

4. _____
_____

3. _____
_____

**Replacing Faulty Genes
with Gene Therapy**

7. _____
_____

6. _____
_____

Cell Differentiation

Cell differentiation is the process by which cells specialize to perform different functions. Multicellular organisms begin as a single cell that divides by mitosis to produce all the cells of the organism, which thus have identical chromosomes. These cells become different from each other because each type of cell uses only the genes it needs. Early in development, cells receive signals that determine their fate by turning on expression of certain genes and silencing others. As development progresses, signals continue to direct subsets of cells, ultimately producing all of the cell types needed by the organism.

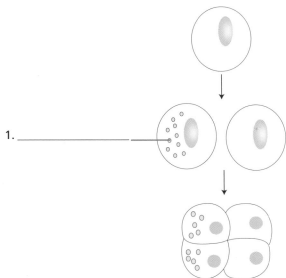

1. _____

How Signals Can Cause Specialization in Cells

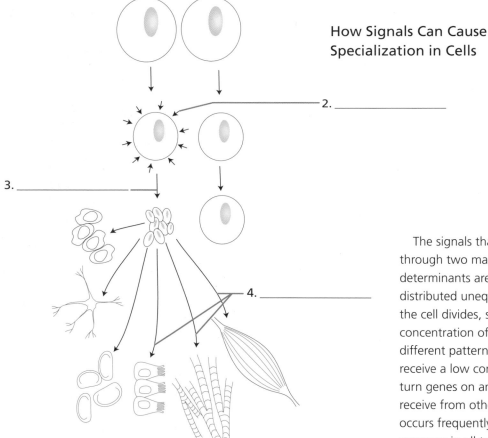

2. _____

3. _____

4. _____

The signals that direct cell differentiation work through two main mechanisms. Cytoplasmic determinants are regulatory molecules that may be distributed unequally in the cytoplasm of the cell. As the cell divides, some of its descendants receive a high concentration of these determinants and will have a different pattern of gene expression than cells that receive a low concentration. Alternatively, cells may turn genes on and off in response to signals they receive from other cells. This process, called induction, occurs frequently throughout development and is common in all types of organisms. Differential gene expression in response to cytoplasmic determinants occurs in insects but is rare in other types of organisms.

Answers

1. cytoplasmic determinants, 2. signals from nearby cells, 3. induction, 4. cell differentiation

Development of Body Plan

The bodies of multicellular organisms can have up to three axes: front/first (anterior) to rear/last (posterior), belly (ventral) to back (dorsal), and left lateral to right lateral. The development of structures in the organism differs based on the location of cells along these axes. The establishment of patterns along these axes, called pattern formation, depends on signaling molecules (morphogens).

Cells produce morphogens, which spread out to form a concentration gradient. Other cells detect the morphogens and respond in different ways depending on the local concentration. In effect, the morphogen concentration tells cells where they are along the gradient axis, allowing cells to differentiate according to their position.

The first morphogen ever discovered is a transcription factor produced by the *bicoid* gene in the fruit fly *Drosophila melanogaster*. Cells in the mother's ovary produce *bicoid* mRNA and transfer it to the egg cell, where it's concentrated in the anterior end. During the multinucleate stage of the embryo, translation of *bicoid* mRNA creates a gradient of the transcription factor. The transcription factor induces different responses in cells depending on its concentration, leading to the establishment of major body plan elements in the embryo—such as the head, thorax, and abdomen—along the anterior-posterior axis. Once established, the body plan persists through development of the larva and into the adult.

Example of Body Plan Development in an Animal (*Drosophila*)

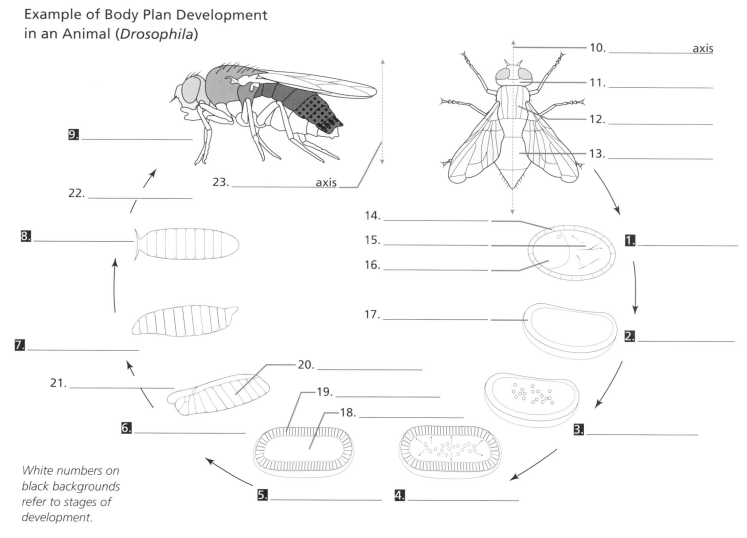

White numbers on black backgrounds refer to stages of development.

Answers

Regulation of Animal Development

The genes that determine major body plan elements work in concert with other regulatory genes to fine-tune this development. Many of these regulatory genes contain the blueprints for transcription factors that control the transcription of other genes. Thus, the activity of one gene turns on another gene, which turns on another gene, and so on, creating a regulatory cascade that specifies developmental events from fertilization to adult.

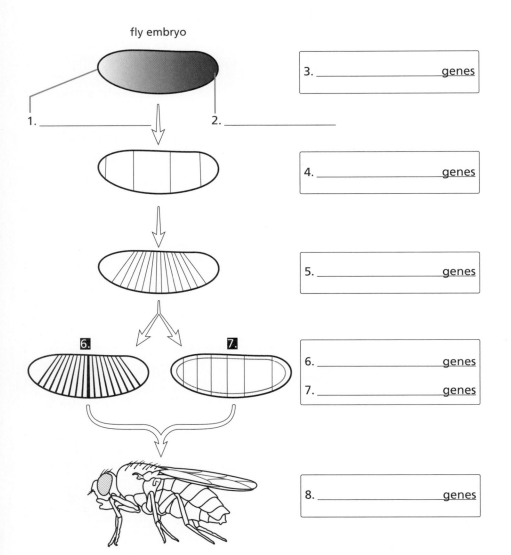

fly embryo

3. _____ **genes**

1. _____   2. _____

4. _____ **genes**

5. _____ **genes**

6.

7.

6. _____ **genes**

7. _____ **genes**

8. _____ **genes**

One of the most well-known regulatory cascades is the one that controls development in *Drosophila melanogaster*. Development begins when morphogens from genes such as *bicoid* establish major body axes. These morphogens act as transcription factors, turning on a set of genes called gap genes, which organize cells into groups of segments along the anterior-posterior axis. In turn, gap genes activate pair-rule genes, which form individual segments within the groups. Pair-rule genes activate segment polarity genes, which fine-tune the anterior-posterior axis within each segment. Gap genes and pair-rule genes also activate homeotic (Hox) genes, which trigger the development of specific structures within each segment by activating effector genes that control cellular differentiation.

Cascade of Genes That Regulate Animal Development

Answers

Homeotic Genes

Homeotic, or Hox, genes are a family of genes found in almost all animals. The genes contain the blueprints for transcription factors, including a special sequence called the homeobox, which determines the shape of the DNA-binding region. Once cells produce the transcription factor from a particular Hox gene, it will recognize and bind to effector genes necessary for proper development of structures in a particular area of the organism. When Hox genes mutate, they cause the development of bizarre animals that have structures in the wrong places, such as legs on the head instead of antennae.

Hox genes from different animals are very similar to each other. They're so similar, in fact, that scientists were able to insert a Hox gene from a mouse into a fruit fly and cause the same effect as if they'd used the fruit fly version of the gene. This close relationship suggests that Hox genes are homologous, meaning they're related because they evolved from a common ancestor. Hox genes have an unusual organization that is also conserved across animal species; the order of the genes along a chromosome aligns with the expression of these genes along the anterior-posterior axis of animals. In other words, Hox genes at one end of the chromosome control development in the anterior end of the organism, while genes at the other end of the chromosome control development in the posterior end.

Organization and Expression of Homeotic (Hox) Genes

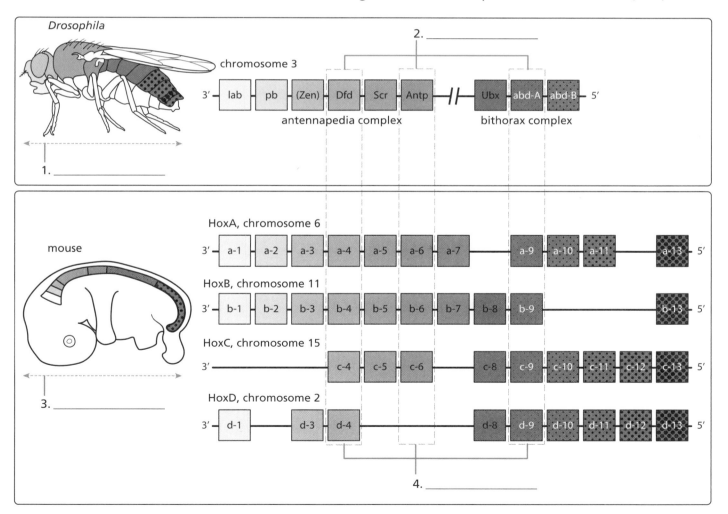

Gastrulation

Gastrulation is one of the most important events during animal development. During gastrulation, cells of the blastula move to reorganize the embryo from a simple hollow ball of cells to a distinctly layered stage called the gastrula. Initially, one layer of the blastula invaginates, or pushes inward, to create two layers. Cells can also create additional layers by moving from a surface layer into the embryo (ingression), or by folding one layer under another (involution). Ultimately, gastrulation creates three primary germ layers called the endoderm, ectoderm, and mesoderm, each of which produces different components of the body. The innermost endoderm produces the internal organs and lining of the gut. The outermost ectoderm gives rise to the nervous system and surface organs such as the skin. The mesoderm in the middle will form the muscular, circulatory, and skeletal systems.

Neurulation occurs after gastrulation as the nervous system develops from cells in the ectoderm. In vertebrates, the portion of the ectoderm adjacent to the notochord thickens to create a raised plate of cells called the neural plate. The plate invaginates, and then seals to create the neural tube, which ultimately becomes the central nervous system. The farthest edges of the neural plate don't become part of the tube; instead, they become neural crest cells that will form pigment cells and the peripheral nervous system.

Gastrulation and Neurulation in a Marine Chordate (Lancelet)

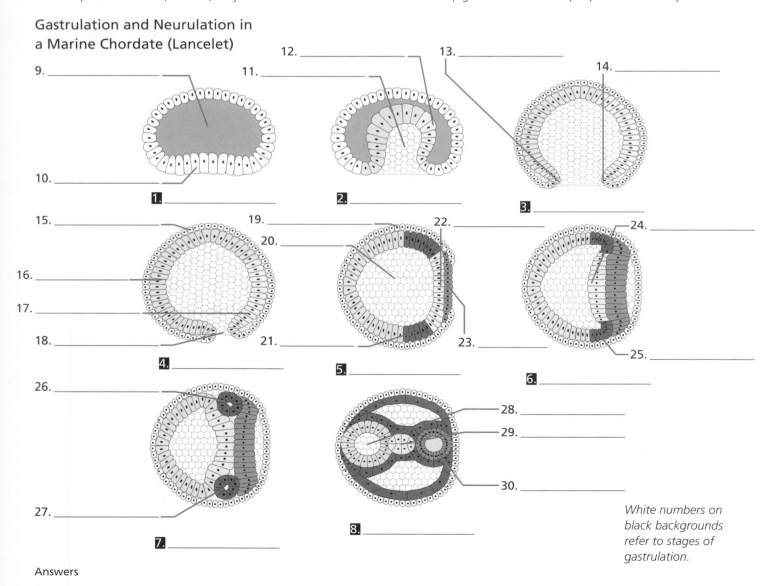

White numbers on black backgrounds refer to stages of gastrulation.

Answers

Embryogenesis in Dicotyledonous Plants

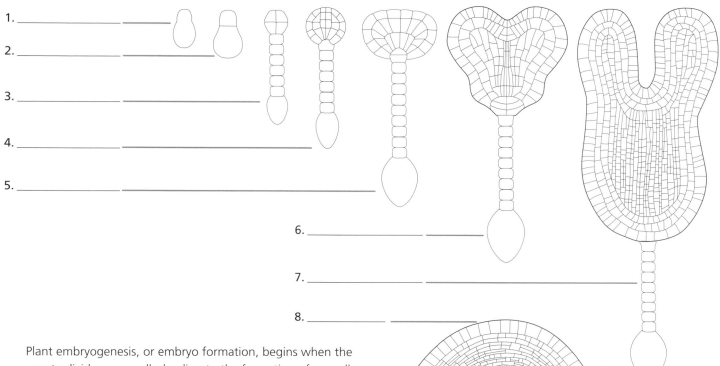

Stages of Embryogenesis (1–8)

1._____ _____
2._____ _____
3._____ _____
4._____ _____
5._____ _____

6._____ _____
7._____ _____
8._____ _____

9._____
10._____

Labels 9 and 10 are structures.

Plant embryogenesis, or embryo formation, begins when the zygote divides unequally, leading to the formation of a small, dense apical cell and a larger vacuolate basal cell. The apical cell divides twice vertically and once horizontally to produce an octant, with clearly divided upper and lower tiers that, respectively, give rise to most of the apical and basal tissues of the seedling. The basal cell divides horizontally to produce the suspensor, most of which acts as supportive tissue. The top cell of the suspensor, the hypophysis, becomes incorporated into the embryo as part of the root meristem. Tangential divisions give rise to the protoderm, a single layer of cells that will become the plant epidermis.

Continued divisions among the inner cells give rise to a globular shape, with cells derived from the upper and lower tiers remaining in alignment along the apical-basal axis as the globular embryo increases in size. Cell divisions in the lower tier create a central region of narrow cells, while cell divisions in the upper tier occur at two opposite tips, leading to a triangular shape and then a heart shape. By the triangular stage, the embryo has formed the primordia of most seedling organs, including the cotyledons, the hypocotyl, and the primary root, as well as basic tissues like provascular tissue, protoderm, and cortex.

Answers

1. fertilized egg, 2. two-cell, 3. octant, 4. globular, 5. triangular, 6. heart, 7. torpedo, 8. mature embryo, 9. cotyledons, 10. vascular tissue

Regulation of Plant Development

Flower formation is a good example of how regulatory cascades direct development in plants. Like in animals, many of the genes in the regulatory cascade are transcription factors that activate or silence other genes. Flower formation begins when signals induce the conversion of the apical meristem to an inflorescence meristem. Next, transcription factors activate floral meristem identity genes, which cause the inflorescence meristem to produce floral meristems and turn on cadastral genes, which delineate regions within the flower. Finally, cadastral genes turn on floral organ identity genes, which direct the formation of floral parts in the correct locations by activating effector genes.

A Model for the Genetic Regulation of Flower Formation in Plants

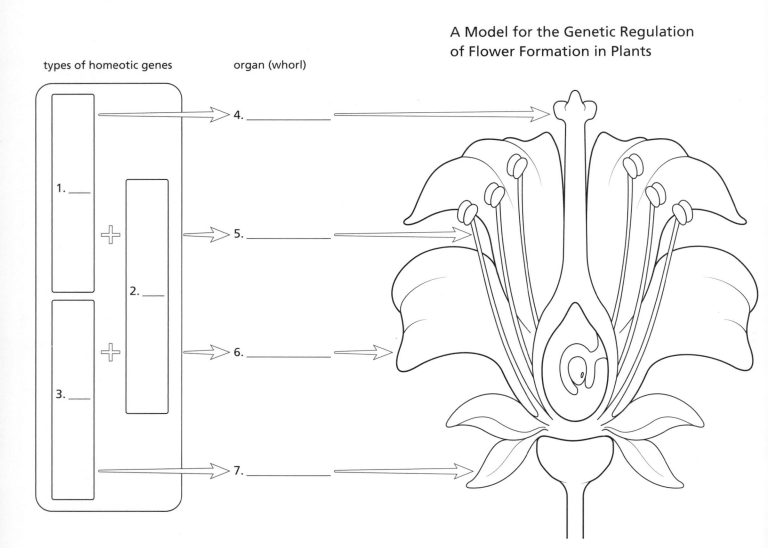

types of homeotic genes   organ (whorl)

1. ___

2. ___

3. ___

4. ___

5. ___

6. ___

7. ___

Floral organ identity genes belong to the MADS-box family of genes. Like Hox genes in animals, MADS-box genes contain the blueprints for transcription factors that are essential to the placement of structures in the correct locations. The MADS box itself is the section of the gene that determines the shape of the DNA binding region. Scientists place floral organ identity genes into categories such as A, B, and C, based on their activity. The ABC model for floral development shows how combinations of genes from these categories direct which organs are formed in each of the four whorls of a flower.

Answers

1. C, 2. B, 3. A, 4. carpels (gynoecium), 5. stamens (androecium), 6. petals (corolla), 7. sepals (calyx)

Homology

Homologous structures are structures that are similar in different species because of their shared ancestry. One of the most well-known examples is that of the bones in vertebrate limbs. From the outside, the flipper of a whale, the paw of a dog or cat, the wing of a bird, and the hand of a human seem to be very different structures that perform very different functions. If you look inside these structures, however, you see that they share the same kinds of bones in the same organization. These apparently different structures have amazing similarity because all these organisms share a common ancestor. In other words, they are all variations from the same original theme.

Other types of homology also provide evidence for the theory of evolution. Genetic homologies, such as those in Hox genes, occur in the DNA of widely different species. Because DNA contains the codes for RNA and protein molecules, genetic homologies can lead to biochemical homologies, such as similarity in enzymes and metabolic pathways. Developmental homologies, such as the temporary presence of a tail in human embryos, also reveal the common ancestry of life on Earth.

1. _____   3. _____   5. _____

2. _____   4. _____   6. _____

Structural Homology of Forelimbs in Different Animal Groups

Label the homologous bones, 1–6.

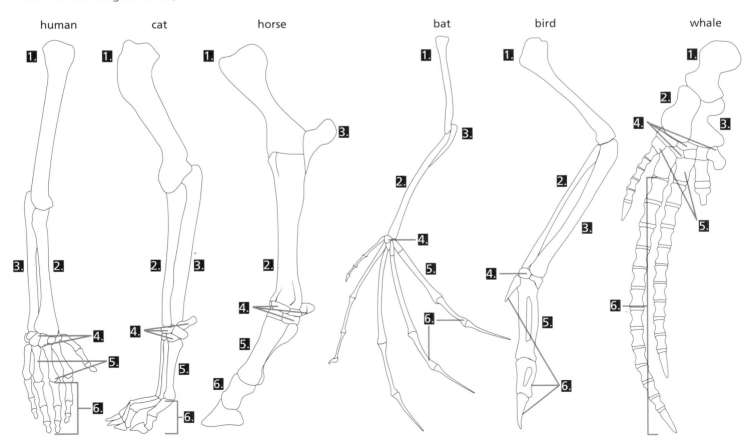

human   cat   horse   bat   bird   whale

Answers

1. humerus, 2. radius, 3. ulna, 4. carpels, 5. metacarpels, 6. phalanges

Natural Selection

Natural selection is a theory first proposed by Charles Darwin that explains one way that populations of organisms might change over time. Darwin included four fundamental principles in his proposal. First, organisms produce more offspring than can be supported by the environment (overproduction). Second, individuals in a population differ in various ways (variation). Third, individuals with traits that favor survival in particular conditions are more likely to survive and produce offspring (natural selection). And fourth, parents pass traits onto offspring, so the next generation will have more favorable traits (adaptations). After many generations, natural selection can lead to the evolution of populations.

Natural selection is commonly referred to as "survival of the fittest." Fittest in this sense, however, is not the same as overall physical fitness. Instead, it refers to biological fitness: the ability of an organism to survive and produce living offspring. What makes an organism biologically fit depends on the selection pressure acting on the population. If the selection pressure is being chased by predators, then the fastest individuals might be the most fit. But if the selection pressure is an attack by a pathogen, then the fittest organisms might actually be those with a genetic defect that somehow makes them resistant to the pathogen.

Process of Natural Selection

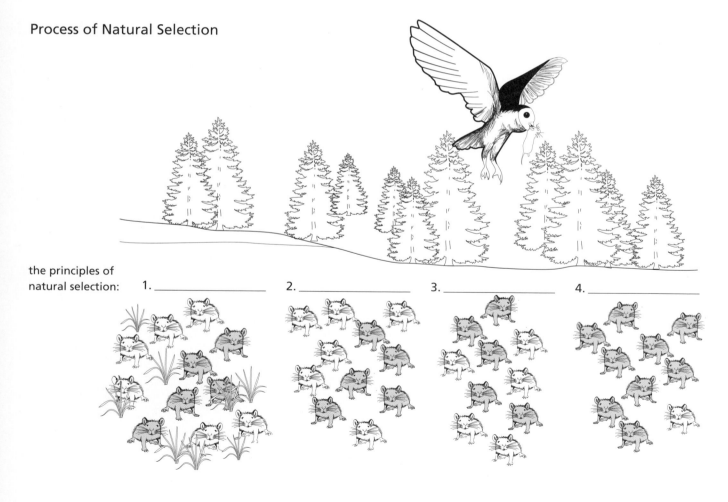

the principles of
natural selection: 1. _____ 2. _____ 3. _____ 4. _____

Answers

1. overproduction, 2. variation, 3. selection, 4. adaptation

Types of Selection

Different types of natural selection increase or decrease the genetic variation of a population by changing the number and frequency of alleles for a particular trait. Stabilizing selection favors individuals with average values of a trait and tends to reduce genetic variation in the population. If human birth weight is too high or too low, the mother or child might not survive. Thus, alleles that contribute to the extremes of this trait are less likely to get passed down to future generations.

Directional selection changes the average value of a trait and tends to reduce genetic diversity of a population. When the Industrial Revolution in England resulted in soot-covered tree trunks, darker peppered moths (*Biston betularia*) were better camouflaged and therefore more likely to survive predation than the lighter moths that used to dominate the population. Over a few generations, directional selection moved the average value of the trait and darker moths dominated the population.

Disruptive selection favors the extreme values of a trait and tends to increase genetic diversity of a population. African mocker swallowtail butterflies (*Papilio dardanus*) mimic the appearance of poisonous butterflies in order to avoid predation. This species is found over a large range, but in each habitat its appearance matches the local poisonous butterfly. In this case, selection is acting to favor different extremes of appearance in each location.

Different Effects Natural Selection Can Have on Populations

Key:
original population   □ □ □ □
population after selection   ────────

type of selection:

1. _____

2. _____

3. _____

nature of selection:

4. _____

5. _____

6. _____

modern horse

ancestral dog-size horse

robins lay medium-sized clutch

small clutches may not have any viable offspring; large clutches may lead to malnourished chicks

light mice have good camouflage on beaches, while dark mice have good camouflage in the forest

medium-shaded mice do not have good camouflage in either environment

Answers

1. directional selection, 2. stabilizing selection, 3. disruptive selection, 4. selection against one extreme, 5. selection against both extremes, 6. selection against the mean

Genetic Drift

Genetic drift refers to changes in genetic variation in a population due to chance events that have no connection to biological fitness. The impact of genetic drift is usually much greater in small populations than in large ones. In very small populations, it can sometimes lead to the loss of alleles for a particular trait or the fixation of alleles when 100 percent of the population has the same allele.

The bottleneck effect is a type of genetic drift that occurs when a species recovers after nearly going extinct. Much of the genetic diversity in the original population is lost as individuals die, leaving only the alleles of the survivors to be passed down to the new generation. This happened when cheetahs (*Acinonyx jubatus*) almost became extinct during the last ice age. The population rebounded, but the genetic diversity is so low that zoos have to be careful when selecting mates for their captive breeding programs in order to avoid inbreeding.

The founder effect is a type of genetic drift that occurs when a few individuals from a population splinter off and start a new population. The alleles of the founding individuals become much more frequent in the new population than they were in the original. The founder effect can happen when storms displace small animals, such as birds or lizards, from their original habitat to new areas.

Mechanisms by which Genetic Drift May Occur

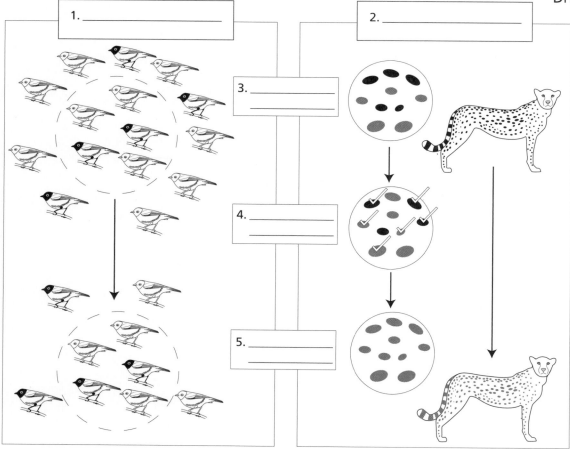

1. _____

2. _____

3. _____

4. _____

5. _____

Answers

Gene Flow

Gene flow occurs when individuals leave one population and join another, taking their alleles with them. The introduction of new alleles to a population can increase its genetic diversity. Although the movement of individuals between populations isn't connected to biological fitness, the resulting movement of alleles can impact the overall fitness of the receiving population. This sometimes happens when animals bred in captivity, such as farmed fish, escape and breed with wild species. Overall, gene flow between two populations tends to make them more similar, or homogenizes them.

Migration and dispersal are two common mechanisms by which gene flow occurs. In migration, organisms move from their original population to a new one. This occurs with organisms that can move freely, such as animals. Dispersal refers to the spreading of gametes or individuals during reproduction. Wind-pollinated plants literally cast their gametes to the wind, sending their alleles far and wide. Plants also use various mechanisms to disperse their seeds, such as wind, water, or animal transport via attachment or ingestion. These dispersal mechanisms can spread individuals and their alleles into new populations.

How Alleles Move Between Populations

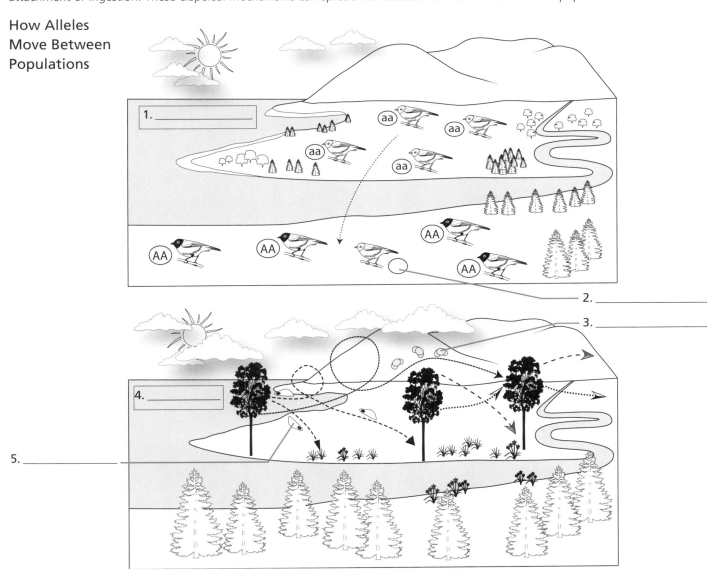

Answers

Speciation

Speciation refers to the evolution of new species from ancestral organisms. To recognize speciation, biologists must be able to identify one species as different from another. For many eukaryotes, biologists apply the biological species concept, which says that if two organisms can produce fertile offspring with each other, then they belong to the same species. For fossil organisms or those that don't reproduce sexually, biologists use the morphospecies concept, which says that if two organisms have the same structures (morphology), then they are the same species. For prokaryotes, biologists also use biochemical and genetic similarity to define species.

The evolution of one species from another can occur when a group of organisms becomes reproductively isolated from the original population. This restricts gene flow between the two populations, allowing them to evolve independently. Allopatric ("other country") speciation occurs when populations become separated by a geographic barrier. In parapatric ("near country") speciation, a species may be spread over a large area. Populations next to each other along the gradient may be able to interbreed, but those at the extremes become too different from each other. In sympatric ("together country") speciation, individuals in the same population evolve independently as they specialize to use certain resources in the habitat.

Mechanisms That Can Lead to Speciation

1. _____

2. _____ speciation   3. _____ speciation   4. _____ speciation

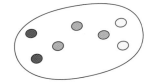

Kaibab and Abert's squirrels on opposites sides of the Grand Canyon

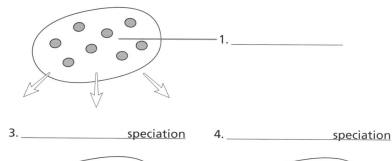

Bullock's and Baltimore orioles in East and West North America

Apple maggot flies on apples and hawthorns

Answers

Phylogeny

Phylogeny is the study of the evolutionary history between populations. Biologists gather morphological, developmental, biochemical, and genetic information about organisms and then build "family trees" called phylogenetic trees to represent the evolutionary relationships inferred from the similarities and differences between traits. The branches of organisms that have more shared characteristics will be closer to each other on the tree, while unique traits, called derived characteristics, trigger branch points on the tree that show when new groups evolved from a common ancestor. Traits that occur in groups of organisms and their common ancestor, but are missing in distant ancestors, are called synapomorphies.

Phylogenetic trees represent evolutionary time, with ancestors positioned toward the roots of the tree and the species being compared at the tips. By tracing the tips of the tree backward to find the branch points, you can identify which species share common ancestors and which ones don't, or which groups shared a common ancestor more recently than others.

An ancestral group plus all of its descendants—and only those descendants—forms a single evolutionary lineage called a monophyletic group (or clade). Biologists use the phylogenetic species concept to define species as the smallest monophyletic groups of the phylogenetic tree that represents all life on Earth.

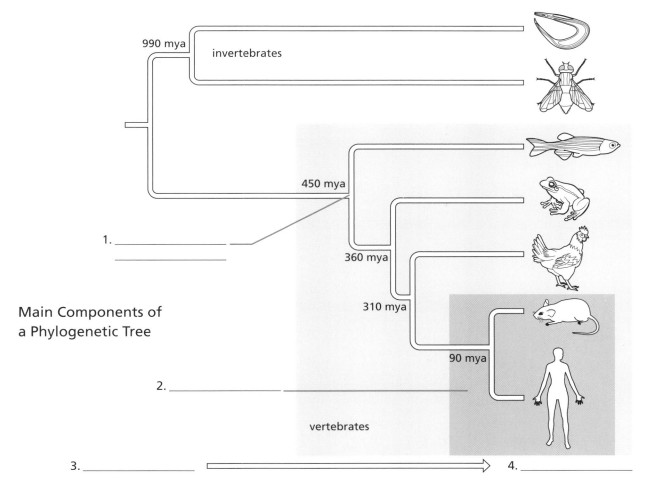

990 mya

invertebrates

450 mya

1. _____
   _____

360 mya

Main Components of a Phylogenetic Tree

310 mya

90 mya

2. _____

vertebrates

3. _____ ⟶ 4. _____

Answers

The Serial Endosymbiotic Theory

The serial endosymbiotic theory (SET) of the origin of eukaryotic cells explains how they acquired mitochondria and chloroplasts by forming permanent associations with bacteria. Symbiosis refers to two organisms living together over long periods of time; the prefix "endo-" means "inside," and the term "serial" indicates that symbiosis happened more than once during the evolution of the eukaryotic cell.

The SET is well supported by multiple lines of evidence. The earliest versions of this idea were based on the morphological similarities between mitochondria and chloroplasts and certain free-living bacteria. Both mitochondria and chloroplasts have a double outer membrane, which supports the idea that the ancestor of the eukaryotic cell caught bacteria in vesicles and kept them as living symbionts. Biochemical analysis shows that ribosomes from chloroplasts and mitochondria have the same structure as ribosomes from bacteria and are different from those in the cytoplasm of eukaryotic cells. Similarly, genetic analysis shows that the genes inside chloroplasts and mitochondria are more similar to genes from bacteria than they are to the genes in the nucleus of the cell. Based on these data, biologists agree that ancestors of the eukaryotic cell engulfed free-living bacteria and that, over time, these symbiotic bacteria evolved to become the mitochondria and chloroplasts of modern eukaryotic cells.

Evolutionary Mechanism That Led to Mitochondria and Chloroplasts

proto-eukaryote or "ancestral prokaryote" (before eukaryote)

1. _____

2. _____
   _____

3. _____ ___

4. _____ ___

5. _____
   _____

6. _____
   _____

7. _____

8. _____

9. _____
   _____

White numbers on black backgrounds refer to cell types or events.

Answers

1. ancestral eukaryote, 2. first symbiotic event, 3. aerobic bacterium, 4. mitochondrion, 5. ancestral heterotrophic eukaryote, 6. second symbiotic event, 7. photosynthetic bacterium, 8. chloroplast, 9. ancestral photosynthetic eukaryote

Bacteria and Archaea

Bacteria and Archaea are two of the three domains of life. In terms of numbers, these prokaryotes represent at least half of all life on Earth. Although their cells appear similar because they both lack nuclei and organelles, more detailed analysis of their fundamental chemistry and genetics reveals that they are quite distinct in their cell wall and membrane composition, as well as in some fundamental aspects of their genetic machinery. Prokaryotes are found in every environment on Earth, including lightless caves, dry deserts, hot springs, and even under the Antarctic ice. Because they perform many essential functions in the environment, such as decomposition, photosynthesis, and nutrient cycling, the rest of life on Earth simply couldn't exist without them.

   Some members of the domain Bacteria are infamous for their ability to cause disease, but the vast majority of bacteria are harmless or even helpful to human life. When viewed through the microscope, they appear in a variety of shapes, including spherical (cocci), rod-shaped (bacillus), and curved (spiral). Most bacteria have a cell wall made of peptidoglycan, which is a target for antibiotics such as penicillin.

   Archaea are known for their ability to live in extreme environments, such as hot springs and very salty (hypersaline) water, but they also live in soil and in freshwater and marine environments. Biochemistry and genetics suggest that Archaea and Eukarya are more closely related to each other than they are to Bacteria.

Some Common Characteristics of Bacteria and Archaea

	Crenarchaeota	Thaumarchaeota	Euryarchaeota
Archaea			

	Actinobacteria	Cyanobacteria	Firmicutes
			3. _____
Bacteria	1. _____	2. _____	
	Spirochetes	Proteobacterium	Chlamydiae
		5. _____	6. _____
	4. _____		7. _____

Answers

1. filaments, 2. heterocyst, 3. bacillus, 4. spiral, 5. flagellum, 6. host cell, 7. elementary body

Close-Up of Nitrogen-fixing Bacteria

One of the most important environmental functions of bacteria is their role in nutrient cycling, such as the nitrogen cycle. All living things need a source of nitrogen atoms in order to build essential molecules, such as proteins and DNA. Nitrogen is potentially abundant in the environment because Earth's atmosphere is approximately 80 percent nitrogen gas (N_2). However, very few organisms can access nitrogen in this form, so in practice, nitrogen availability is one of the most limiting factors in natural environments.

How Nitrogen-Fixing Bacteria Colonize Plant Roots

Nitrogen-fixing bacteria play an essential role for other organisms because they can capture nitrogen gas from the atmosphere and use it to build nitrogen-containing molecules. Some nitrogen-fixing bacteria, such as cyanobacteria and *Azotobacter*, live freely in soil and water. However, one of the most important groups, the rhizobia, must live symbiotically with plant roots in order to fix nitrogen.

Rhizobia from the soil enter the roots of plants in the bean family (legumes). The bacteria initially attach to root hairs, causing them to curl inward and form tubules called infection threads. The bacteria then multiply inside these threads as they grow and move into the root tissue. Once there, the bacteria colonize the root cells, multiplying and changing their morphology into bacteroids. The expansion of the root cells forms nodules on the roots of the legumes.

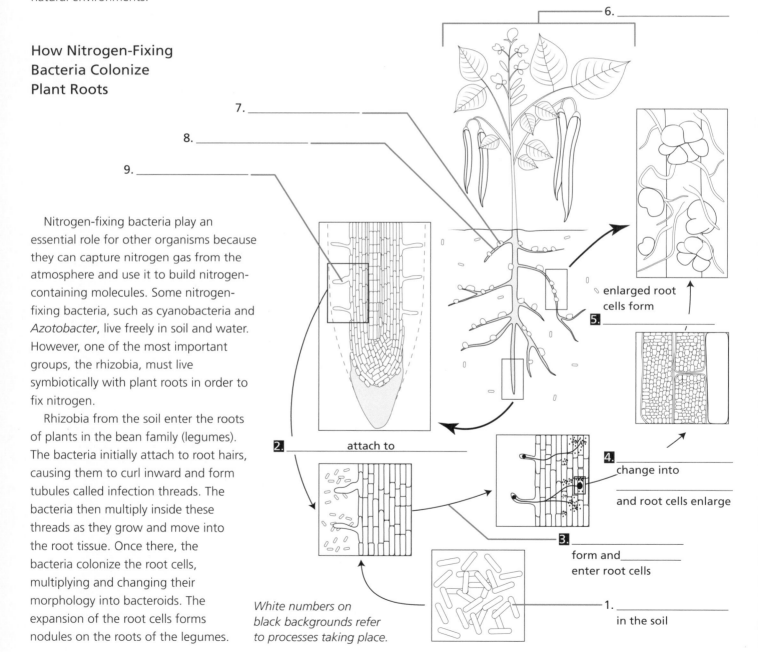

6. _____

7. _____

8. _____

9. _____

enlarged root cells form

5. _____

2. _____ attach to _____

4. _____ change into _____ and root cells enlarge

3. _____ form and_____ enter root cells

1. _____ in the soil

White numbers on black backgrounds refer to processes taking place.

Answers

1. free-living rhizobia, 2. rhizobia, 3. infection threads, 4. bacteria, bacteroids, 5. nodules, 6. legume plant, 7. nodule, 8. root, 9. root hair

Protista

The Protista contains all of the eukaryotic organisms except for plants, fungi, and animals. Once considered a kingdom, it is now known to be an extremely diverse group containing several distinct lineages that could be considered kingdoms in their own right. Protists are mostly microscopic organisms that perform a wide variety of functions in ecosystems. They are abundant in aquatic habitats, where they are important components of food webs. Some protists are photosynthetic, and a few are pathogenic. Although the exact phylogeny of protists is still being studied, it's clear that animals, plants, and fungi all evolved from protist ancestors.

Genetic analysis reveals at least seven lineages within the eukaryotes. The Amoebozoa includes amoebae and slime molds, both of which lack cell walls. The Excavata includes parabasalids, diplomonads, euglenids, and kinetoplastids. These asymmetrical cells have a feeding groove on one side and unusual mitochondria. The Rhizaria includes actinopods and foraminiferans, which form beautiful outer coverings made of materials like silica and calcium carbonate. The Alveolata consists of ciliates, dinoflagellates, and apicomplexans, all of which have small vesicles called alveoli under their plasma membranes. The Stramenopila have a similar flagellar structure and include the water molds, diatoms, and brown algae.

Two eukaryotic lineages contain organisms that biologists don't traditionally consider to be protists. The red algae and glaucophyte algae form a lineage called the Archaeplastida with the green algae and land plants. All of these organisms have chloroplasts surrounded by a double membrane. The Opisthokonta includes both the fungi and animal kingdoms, which have cells with a similar flagellar structure and flat cristae in their mitochondria.

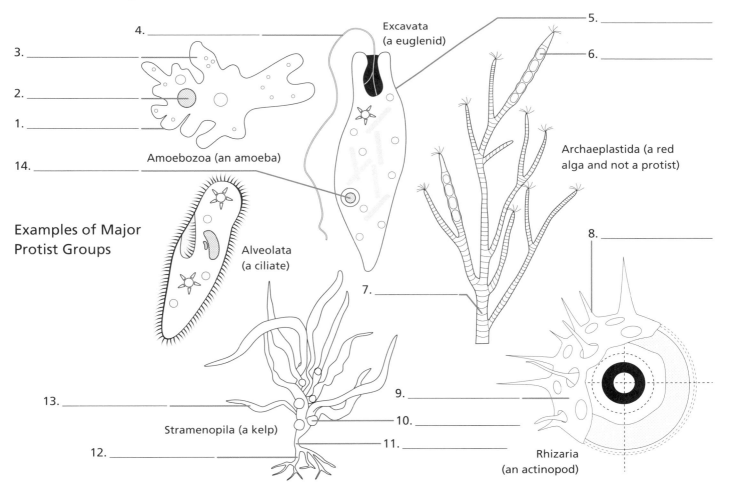

4. _____

Excavata
(a euglenid)

5. _____

3. _____

6. _____

2. _____

1. _____

Amoebozoa (an amoeba)

Archaeplastida (a red alga and not a protist)

14. _____

8. _____

Examples of Major Protist Groups

Alveolata
(a ciliate)

7. _____

13. _____

9. _____

10. _____

Stramenopila (a kelp)

11. _____

12. _____

Rhizaria
(an actinopod)

Answers

1. plasma membrane, 2. nucleus, 3. pseudopod, 4. flagellum, 5. pellicle, 6. tetraspore, 7. thallus, 8. spine, 9. cortical shell, 10. pneumatocyst, 11. stipe, 12. holdfast, 13. blade, 14. nucleus

Major Plant Groups

Modern genetic analysis shows that green algae and land plants form a monophyletic group. Biologists commonly identify this group as the kingdom Plantae within the domain Eukarya. The plants in this kingdom are photosynthetic and have chloroplasts that contain the pigments chlorophyll *a*, chlorophyll *b*, and α-carotene.

Green algae are abundant in freshwater and nearshore marine environments, where they are an important part of food webs. Some green algae are single-celled and microscopic, while others, such as the sea lettuce (*Ulva lactuca*), are multicellular. The ancestor of the land plants was a member of this group.

The most ancient group of land plants, the nonvascular plants, don't contain specialized cells to conduct water and sugar. They include mosses, liverworts, and hornworts. Although these plants do have adaptations to drier environments, such as thick-walled spores and embryos that remain attached and supported by the parent plant, they need moist environments and require water for reproduction. The seedless vascular plants include the club mosses, whisk ferns, ferns, and horsetails. These plants can survive in drier environments due to their vascular tissue but still require water for reproduction.

The seed plants have even better adaptations to life on land. The evolution of seeds provided protection for embryos, and pollen allows distribution of male gametes without water. Gymnosperms such as ginkgoes, pines, cycads, and redwoods produce their seeds in cones. Angiosperms include all the flowering plants, which produce their seeds in fruits.

Examples from the Major Plant Groups

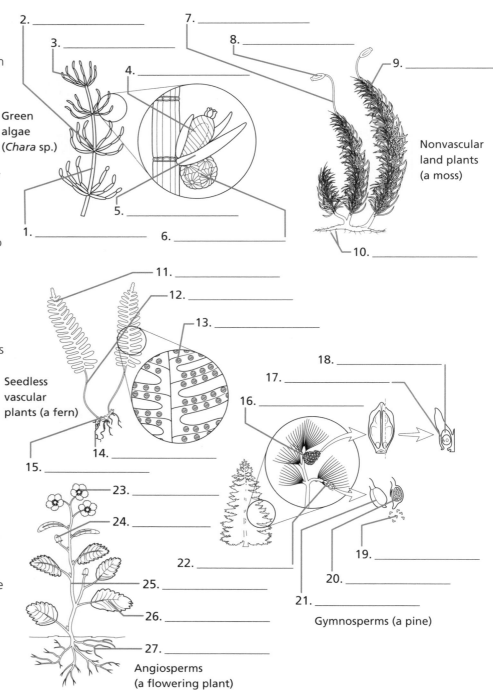

2. _____

3. _____

4. _____

Green algae (*Chara* sp.)

5. _____

1. _____

6. _____

7. _____

8. _____

9. _____

Nonvascular land plants (a moss)

10. _____

11. _____

12. _____

13. _____

Seedless vascular plants (a fern)

14. _____

15. _____

16. _____

17. _____

18. _____

19. _____

20. _____

21. _____

22. _____

Gymnosperms (a pine)

23. _____

24. _____

25. _____

26. _____

27. _____

Angiosperms (a flowering plant)

Major Fungal Groups

Fungi may appear plant-like, but they are a distinct group of heterotrophic organisms that are characterized by absorptive nutrition and chitin-containing cell walls. Fungi typically grow as long chains of cells called filaments that spread throughout their environment, forming a fungal web called a mycelium. Fungi reproduce by forming spores, and many species reproduce both sexually and asexually. Along with the bacteria, fungi are primary decomposers, playing an essential role in the environment by recycling the nutrients available in dead organisms.

Fungi are divided into groups based on their genetics and reproductive structures. The chytrids live in aquatic or moist environments. They don't produce true mycelia; instead, they grow as tiny spherical structures on decaying organic material. They are also distinct from the other fungi in that they produce swimming spores called zoospores. The zygomycetes include the familiar molds that grow on food. They can reproduce asexually by producing mitospores in cells called mitosporangia, or sexually by producing zygospores when hyphae of opposite mating types meet. The ascomycetes include the morels, the antibiotic-producing *Penicillium*, and the yeasts. When they reproduce sexually, they form spores in sac-like structures called asci. Asexual reproduction frequently occurs by the production of broom-like structures called conidia. The basidiomycetes include the mushrooms, shelf fungi, puffballs, rusts, and smuts. They reproduce sexually by producing spores on a club-shaped structure called a basidium.

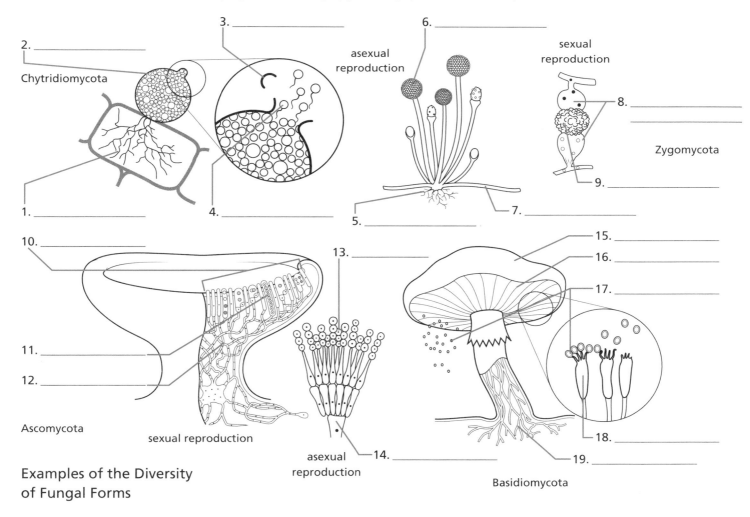

Examples of the Diversity of Fungal Forms

Close-Up of Mycorrhizae

Most plants form mycorrhizae, which are symbiotic associations between their roots and soil fungi. The relationship between the two organisms is mutualistic: the fungus receives food from the photosynthesis of the plant, while the plant gains additional water and minerals from the fungus. Mycorrhizae can spread over large areas, effectively extending the absorption zone around a plant by a factor of hundreds to thousands. Mycorrhizae also help spread signaling molecules between plants, which communicate messages about the presence of pathogens or insects and can increase the production of defensive compounds in plants.

In some mycorrhizae, the fungal partner stays on the outside of the root, while in others the fungus penetrates and grows inside the root cells. In ectomycorrhizae, the fungus forms a sheath (called a mantle) over the surface of the root. Although some hyphae penetrate between cells of the root to form a network called a Hartig net, the hyphae don't actually enter plant cells. In endomycorrhizae, the fungal hyphae actually penetrate the cell walls of root cells and grow in between the wall and the plasma membrane, pushing the plasma membrane inward as they grow. Some endomycorrhizae form branched structures called arbuscles, which increase the contact between the hyphae and the plant membranes. The increased surface contact facilitates the transfer of materials between the two partners.

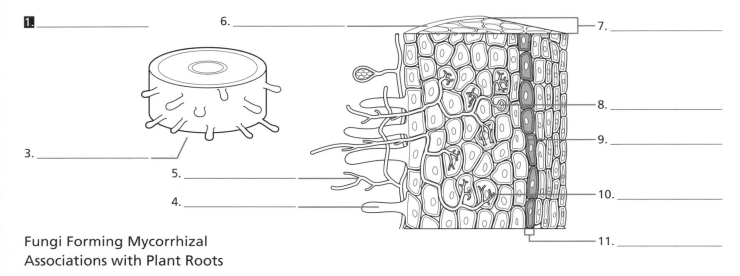

1. _____
3. _____
4. _____
5. _____
6. _____
7. _____
8. _____
9. _____
10. _____
11. _____

Fungi Forming Mycorrhizal Associations with Plant Roots

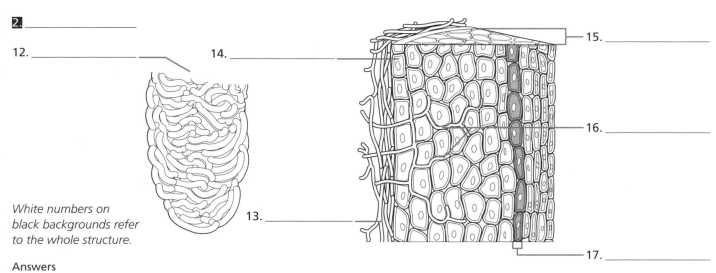

2. _____
12. _____
13. _____
14. _____
15. _____
16. _____
17. _____

White numbers on black backgrounds refer to the whole structure.

Answers

1. endomycorrhizae, 2. ectomycorrhizae, 3. plant root, 4. root hair, 5. hyphae, 6. epidermis, 7. cortex, 8. vesicle, 9. plasma membrane, 10. arbuscle, 11. endodermis, 12. plant root, 13. mantle (fungal sheath), 14. epidermis, 15. cortex, 16. hyphae (Hartig net), 17. endodermis

Protostome Animals

Animals are a large group of multicellular, heterotrophic organisms that lack cell walls and form a hollow ball of cells called a blastula early in development. Most animals have nerves and muscles, although these are lacking or rudimentary in the most ancient groups of animals: sponges, ctenophores (comb jellies), and cnidarians (jellies, corals, and anemones). While radial symmetry is typical for these ancient groups, bilateral symmetry is common to most animals.

Based on genetic analysis, scientists divide the bilaterally symmetrical animals into two major groups: the protostomes and the deuterostomes. The names of the two groups are derived from the fate of the initial pore (blastopore) produced early in gastrulation (see p. 86): in deuterostomes ("second mouth"), this pore becomes the anus, and the mouth forms later; in protostomes ("first mouth"), the pore may become the mouth, the anus, both, or neither.

Genetic analysis further divides the protostomes into two lineages called lophotrochozoans and ecdysozoans. The lophotrochozoans includes rotifers, flatworms, annelids, and mollusks. Scientists define these animals by their ability to grow continuously and by the presence in some members of a feeding structure called a lophophore or a particular type of larva called a trochophore. Ecdysozoans include roundworms, water bears, velvet worms, and arthropods. The organisms in this group are united by their intermittent growth pattern and the need to shed, or molt, their outer covering periodically.

Characteristics of Protostome Animals and Examples of Major Groups

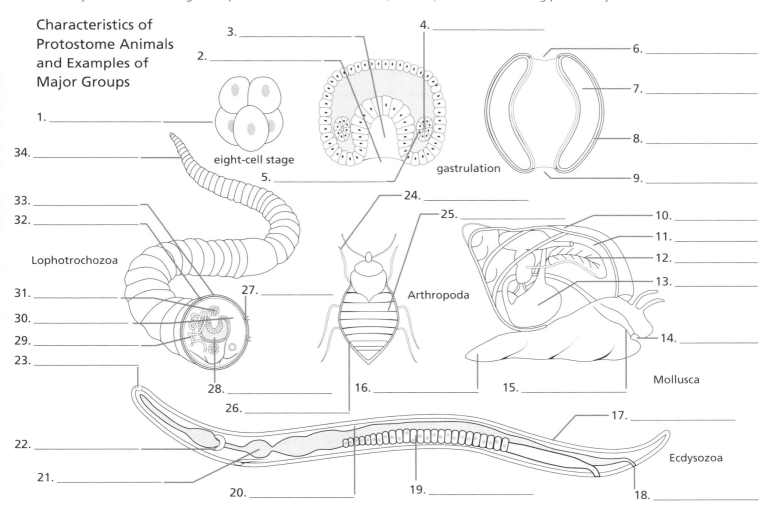

eight-cell stage

gastrulation

Lophotrochozoa

Arthropoda

Mollusca

Ecdysozoa

1. _____
2. _____
3. _____
4. _____
5. _____
6. _____
7. _____
8. _____
9. _____
10. _____
11. _____
12. _____
13. _____
14. _____
15. _____
16. _____
17. _____
18. _____
19. _____
20. _____
21. _____
22. _____
23. _____
24. _____
25. _____
26. _____
27. _____
28. _____
29. _____
30. _____
31. _____
32. _____
33. _____
34. _____

Answers

1. spiral cleavage, 2. blastopore, 3. archenteron, 4. coelom, 5. mesoderm, 6. anus, 7. coelom, 8. mesoderm, 9. mouth, 10. shell, 11. mantle, 12. gill, 13. stomach, 14. mouth, 15. radula, 16. foot, 17. cuticle, 18. anus, 19. ovary, 20. pseudocoelom, 21. intestine, 22. nerve ring, 23. mouth, 24. jointed appendage, 25. segmented body, 26. external skeleton, 27. setae, 28. intestine, 29. nephridium, 30. coelom, 31. dorsal blood vessel, 32. longitudinal muscle, 33. circular muscle, 34. cuticle

Deuterostome Animals

The deuterostomes contain four lineages of animals: echinoderms, hemichordates, xenoturbellids, and chordates. Echinoderms include sea stars, sea urchins, and sea cucumbers. Although the larvae of these animals are bilaterally symmetrical, their adult forms all have a unique five-sided form of radial symmetry. Echinoderms are also united by the presence of a calcium carbonate endoskeleton just under their epidermis and a system of fluid-filled tubes called the water vascular system, which they use to move, breathe, and circulate food and waste.

Chordates are animals that display four key characteristics at some time in their life cycle: pharyngeal slits or pouches; a dorsal hollow nerve cord; a supportive, flexible rod called a notochord that runs the length of the body; and a muscular post-anal tail. In chordates called vertebrates, cartilaginous or bony structures protect the brain and the nervous tissue along the dorsal side of the body. Scientists call animals that lack these hard structures invertebrates.

Both the hemichordates and xenoturbellids are relatively small groups of animals. Hemichordates, which include the burrowing acorn worms, got their name because they share some characteristics with the chordates. In particular, they have gill slits, a structure similar to a notochord, and a dorsal nerve cord. Xenoturbellids are small, ciliated worm-like animals that consist of only one genus with two species.

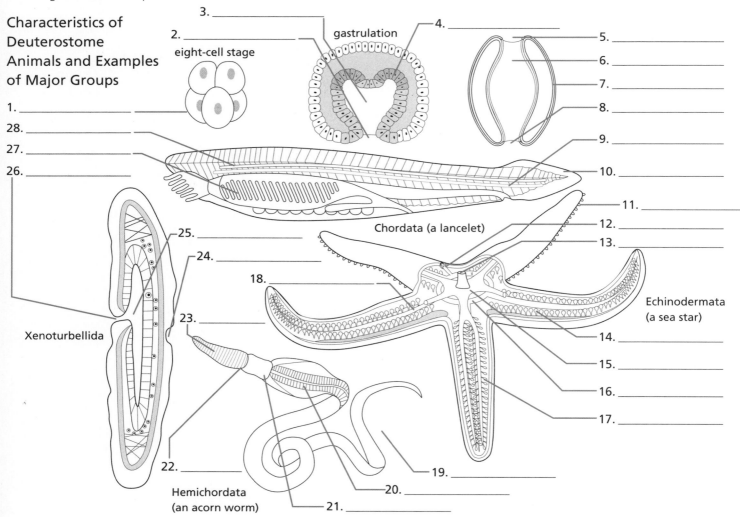

Characteristics of Deuterostome Animals and Examples of Major Groups

3. _____

2. _____
eight-cell stage

gastrulation

4. _____

5. _____

6. _____

7. _____

8. _____

9. _____

1. _____

28. _____

27. _____

26. _____

10. _____

11. _____

12. _____

13. _____

Chordata (a lancelet)

25. _____

24. _____

18. _____

23. _____

Echinodermata (a sea star)

14. _____

15. _____

16. _____

17. _____

Xenoturbellida

22. _____

Hemichordata (an acorn worm)

21. _____

19. _____

20. _____

Answers

Anatomy of a Plant

Basic Structures of a Plant

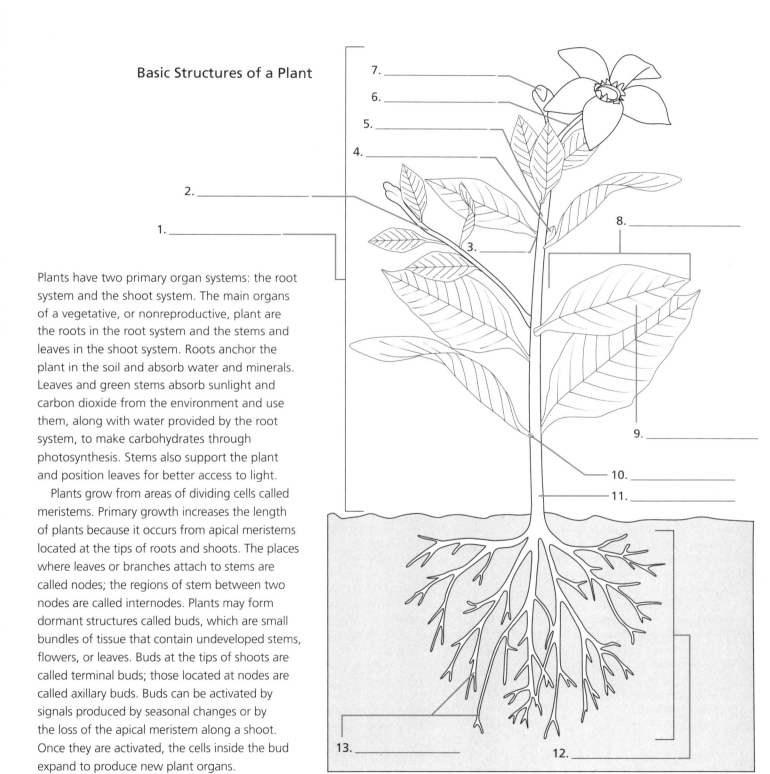

Plants have two primary organ systems: the root system and the shoot system. The main organs of a vegetative, or nonreproductive, plant are the roots in the root system and the stems and leaves in the shoot system. Roots anchor the plant in the soil and absorb water and minerals. Leaves and green stems absorb sunlight and carbon dioxide from the environment and use them, along with water provided by the root system, to make carbohydrates through photosynthesis. Stems also support the plant and position leaves for better access to light.

Plants grow from areas of dividing cells called meristems. Primary growth increases the length of plants because it occurs from apical meristems located at the tips of roots and shoots. The places where leaves or branches attach to stems are called nodes; the regions of stem between two nodes are called internodes. Plants may form dormant structures called buds, which are small bundles of tissue that contain undeveloped stems, flowers, or leaves. Buds at the tips of shoots are called terminal buds; those located at nodes are called axillary buds. Buds can be activated by signals produced by seasonal changes or by the loss of the apical meristem along a shoot. Once they are activated, the cells inside the bud expand to produce new plant organs.

Answers

Plant Tissues

Three types of tissue make up the organs of plants: dermal tissue, vascular tissue, and ground tissue. Dermal tissue forms a protective layer on exterior plant surfaces. The epidermis is a single layer of cells that lack chloroplasts and secrete waxy materials, forming a waterproof outer layer called the cuticle.

Vascular tissue conducts materials between the shoot and root systems of the plant. Xylem transports water and minerals from the roots throughout the plant body. Phloem moves sugars from photosynthesis in the shoots to all the plant cells. Xylem and phloem group together in vascular bundles that run throughout the plant, much like veins and arteries spread through animal bodies.

Ground tissue is the supportive tissue that makes up the rest of plant organs between the dermal tissue and the vascular tissue. Scientists divide the cells that make up ground tissue into three categories. Parenchyma cells have thin walls and remain alive at maturity. Collenchyma cells provide structural support due to localized areas of thickening in their cell walls. Sclerenchyma cells form thick cell walls that contain lignin, a rigid molecule also found in woody plants. These cells often die at maturity, but their cell walls remain to reinforce plant tissues.

Basic Tissues Found in Plants

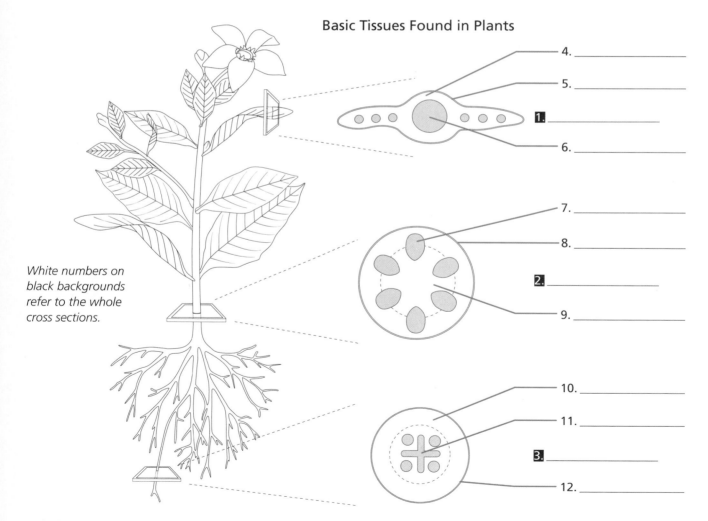

White numbers on black backgrounds refer to the whole cross sections.

4. _____
5. _____
1. _____
6. _____

7. _____
8. _____
2. _____
9. _____

10. _____
11. _____
3. _____
12. _____

Answers

1. leaf, 2. stem, 3. root, 4. ground tissue, 5. dermal tissue, 6. vascular tissue, 7. vascular tissue, 8. dermal tissue, 9. ground tissue, 10. ground tissue, 11. vascular tissue, 12. dermal tissue

Leaf Structure

Leaves are plant organs specialized for photosynthesis. They're often flattened structures that plants can space and angle for maximum light absorption, but they may also be needle-shaped structures. In broad leaves, the flat blade of the leaf is attached to the plant by the petiole. A waxy cuticle covers the upper and lower epidermal surfaces, helping to prevent water loss from the leaf. Openings called stomata (or stomates) allow carbon dioxide to enter and oxygen to exit through the lower epidermis. Guard cells on either side of each stoma swell and relax to open and close the opening. Vascular tissue runs through the leaf in veins, bringing the water needed for photosynthesis through the xylem and carrying away sugars via the phloem. Bundle sheath cells surround the vascular tissue.

In broad-leaved plants, two types of parenchyma cells make up the mesophyll, or interior of the leaf. Both types of cells are packed with chloroplasts because their primary function is photosynthesis. Tall, columnar palisade parenchyma cells form a row under the upper epidermis. Loosely packed cells form the spongy parenchyma below the palisade layer. The air spaces around the cells in the spongy parenchyma facilitate the gas exchange necessary for photosynthesis.

Major Structural Elements of a Leaf

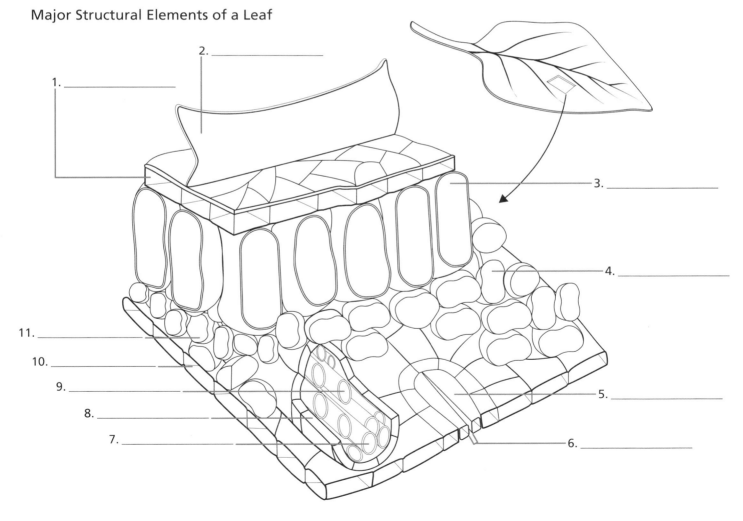

1. _____
2. _____
3. _____
4. _____
5. _____
6. _____
7. _____
8. _____
9. _____
10. _____
11. _____

Answers

Close-Up of Stomata

Plants must balance their need to conserve water with their need to exchange gases like carbon dioxide and oxygen with the environment. Many plants have waxy cuticles on top of their epidermal surfaces, which conserve water but don't allow gas exchange. Plants solve this problem by having surface openings called stomata that allow gases to move into and out of the leaf. Carbon dioxide flows into the air spaces in the spongy mesophyll, where it can be easily absorbed by mesophyll cells for photosynthesis.

When light is available and plants have adequate water, the guard cells surrounding the stomata swell up, becoming turgid and opening the stomata. First, blue light receptors in the guard cells detect light and trigger the active transport of hydrogen ions (H^+) out of the cell, creating a negative electrical potential inside the cell. Next, voltage-gated potassium channels open, and potassium ions (K^+) flow into the guard cells. Other ions follow, increasing the solute concentration and causing water to flow in by osmosis.

In response to water stress, the plant hormone abscisic acid triggers the movement of ions such as chloride (Cl^-) out of the guard cells, stopping any further uptake of K^+. As the solute concentration decreases, water flows out of the cells, which become flaccid, thus closing the stoma.

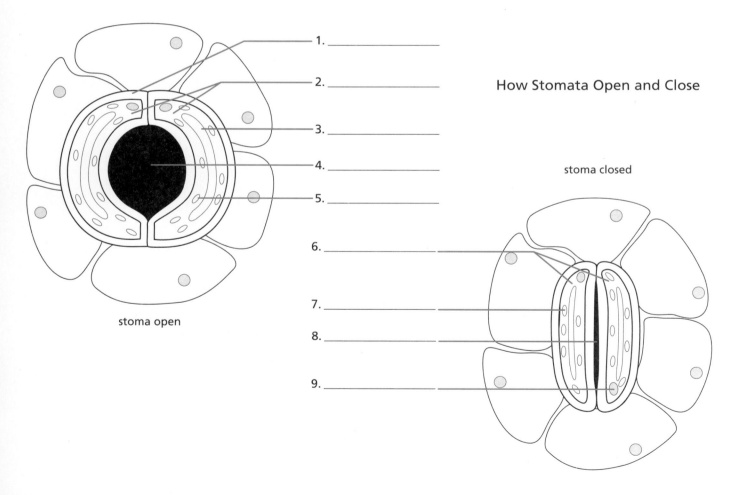

1. _____

2. _____

3. _____

4. _____

5. _____

6. _____

7. _____

8. _____

9. _____

stoma open

How Stomata Open and Close

stoma closed

Answers

1. cell wall, 2. guard cells (swollen), 3. vacuole, 4. stoma, 5. chloroplast, 6. guard cells (flaccid), 7. chloroplast, 8. stoma, 9. nucleus

Plant Stems

Plant stems support and elevate plant organs such as leaves, flowers, and fruits. They also contain vascular tissue to conduct water, minerals, and sugar throughout the plant. Scientists make cross sections of plant stems to study the types and organization of tissues within the stem. In herbaceous (non-woody) stems, the outermost layer of cells is the epidermis. In cross sections, vascular tissue appears as vascular bundles containing xylem toward the inner part of the stem and phloem toward the outside. Thick-walled sclerenchyma cells called fibers often cluster adjacent to the phloem cells, providing support. Ground tissue fills the stem around the vascular bundles.

   The stems of flowering plants show two very different types of organization. In monocots—plants that have one cotyledon (embryonic leaf) in their seeds, such as corn and other grasses—the vascular bundles are scattered throughout the stem. In dicots—plants that have two cotyledons in their seeds, such as beans, roses, and asters—the vascular bundles form a ring around the outer edge of the stem. The ground tissue around the bundles is the cortex, and the ground tissue in the interior of the stem is the pith. In dicots that produce wood through secondary growth, a narrow band of cells separates the xylem from the phloem in the vascular bundles. These cells become the vascular cambium, which divides to produce new vascular tissue during secondary growth.

Organization of Tissues in Plant Stems

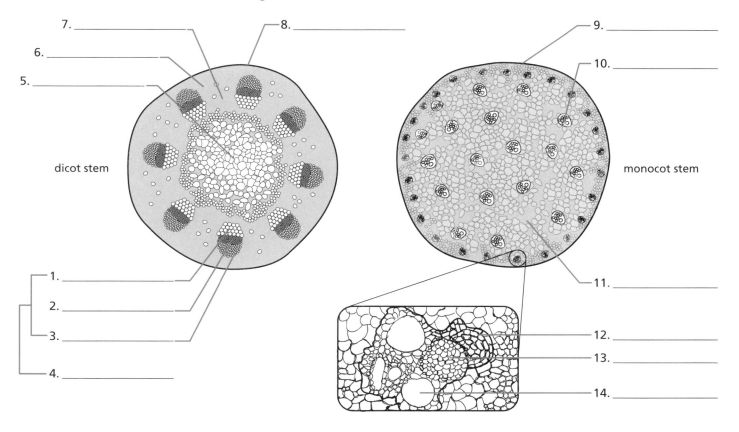

dicot stem

monocot stem

7. _____
6. _____
5. _____
8. _____
9. _____
10. _____
1. _____
2. _____
3. _____
4. _____
11. _____
12. _____
13. _____
14. _____

Answers

Close-Up of Xylem and Phloem

The two types of water-conducting cells in xylem are tracheids and vessel elements. These cells have thickened cell walls with thinner areas called pits, which allow the flow of water between cells. Both cell types are dead at maturity. Tracheids, which appear first in the fossil record, are long, thin cells with tapered ends. Vessel elements, found primarily in flowering plants, are shorter and wider than tracheids. The ends of vessel elements, called perforation plates, connect the elements into columns. These plates have various patterns of openings that allow water to flow freely from one vessel element to the next. Xylem also contains supportive xylem fibers and xylem parenchyma cells.

The sugar-conducting cells in phloem are called sieve tube elements. As they develop, these cells lose many of their organelles, including their nuclei. The sieve elements form a partnership with adjacent cells called companion cells, which supply their proteins and help load them with sugars. Sieve elements directly connect with their companion cells and each other via plasmodesmata. The individual elements connect into sieve tubes via their end walls, called sieve plates. Sieve plates have large holes in them to allow sugar to flow freely through the sieve tubes. Phloem also contains phloem fibers and phloem parenchyma cells.

Details of the Cell Types Found in Xylem and Phloem

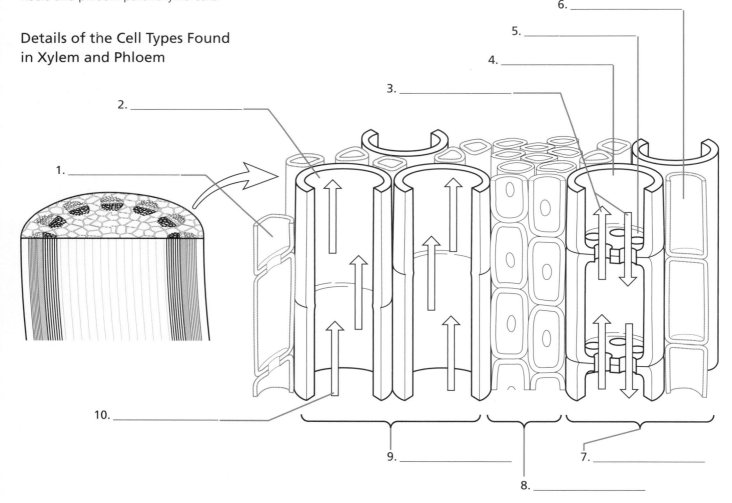

1. _____
2. _____
3. _____
4. _____
5. _____
6. _____
7. _____
8. _____
9. _____
10. _____

Answers

Secondary Growth

Secondary growth increases the diameter of plants that form woody tissue. This growth relies on the action of two lateral meristems, called the vascular cambium and the cork cambium. In older areas of the plant, primary growth stops and plant hormones activate cells in the vascular bundles to differentiate into the vascular cambium. The vascular cambium divides, producing secondary xylem tissue toward the inside of the stem and secondary phloem toward the outside. Changes in the diameter of tracheids and vessels in response to seasonal rainfall can create annual growth rings in the wood of some plants. As the stem expands, older layers of the more delicate cells in the phloem get stretched and broken, leaving a narrow ring of the most recent secondary phloem underneath the cork layer of the stem.

The original epidermis also breaks as the stem widens. Parenchyma cells in the cortex differentiate into the cork cambium, producing cork cells, which are dead at maturity. Together, the cork cells and cork cambium, plus a thin layer of parenchyma cells underneath the cork (phelloderm), form a layer called the periderm, which makes up the outer bark of woody stems. As the stem continues to grow, a new cork cambium forms underneath the old cambium so that the periderm can be continually replenished.

Process of Secondary Growth

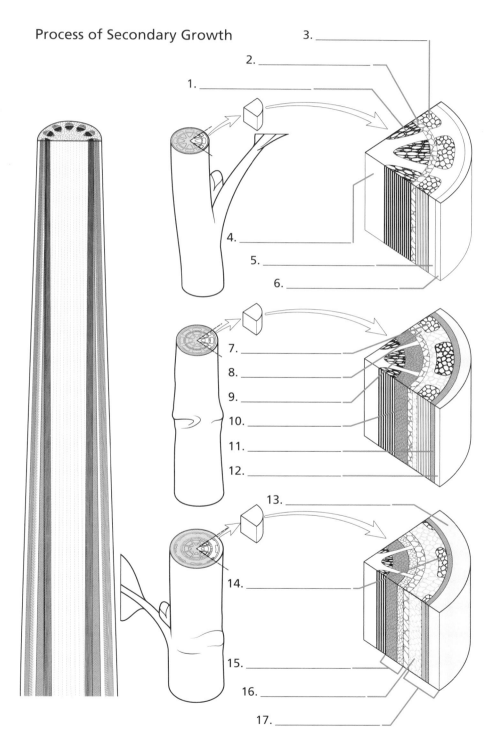

1. _____
2. _____
3. _____
4. _____
5. _____
6. _____
7. _____
8. _____
9. _____
10. _____
11. _____
12. _____
13. _____
14. _____
15. _____
16. _____
17. _____

Answers

1. primary xylem, 2. vascular cambium, 3. primary phloem, 4. pith, 5. cortex, 6. epidermis, 7. primary phloem, 8. secondary phloem, 9. primary xylem, 10. secondary xylem, 11. cork cambium, 12. cork, 13. cork, 14. cork cambium, 15. secondary xylem, 16. secondary phloem, 17. bark

Transpiration

Transpiration is the evaporation of water from plants, primarily via stomata in the leaves. It is important to plants because it provides the force to pull water up through the xylem. Water molecules tend to stick to each other (cohesion) due to the attraction of hydrogen bonds. They also stick to the cell walls in the xylem (adhesion). As water exits the leaves, the resulting negative pressure (tension) causes water to flow up through the xylem. Water enters the xylem as it flows from the soil into the roots. This process of moving water up and out of a plant, called the cohesion-tension theory, is very similar to what happens when a person sucks water through a drinking straw.

The rate of transpiration and water movement is affected by environmental conditions and plant structures. Dry or windy conditions increase the rate of transpiration and water movement. Plant structures such as hairs on the lower surface of leaves can create a sheltered boundary layer with higher humidity that slows transpiration. When rates of transpiration become too strong, cavitation in the xylem interrupts water flow. Plants can repair this situation at night by actively loading solutes into their roots to increase water uptake by osmosis. The increased root pressure drives the water column up through the xylem and removes the gas bubbles.

The Process of Transpiration

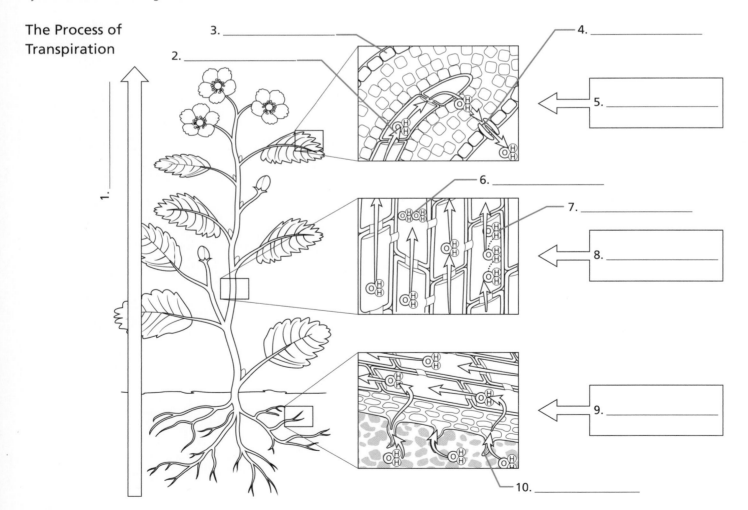

3. _____

2. _____

4. _____

5. _____

1. _____

6. _____

7. _____

8. _____

9. _____

10. _____

Answers

Root Systems

Roots absorb water and minerals and typically anchor plants in soil. Identifying a plant structure as a root can be challenging, however, because some roots grow above the ground and some stems grow underground. Also, both stems and roots can be modified for food storage and appear very similar, such as potatoes (stems) and turnips (roots).

True roots can be recognized by their anatomical features and the fact that they don't produce stems or leaves. Layers of mucilaginous cells form a protective root cap over the tips of growing roots, which contain their apical meristems. Behind the root cap, the root epidermis produces root hairs to create more surface area for absorption. The root cortex is between the epidermis and an inner endodermis. Water flows into the root epidermal cells by osmosis and then passes between the cells of the cortex until it reaches a waxy layer of cell walls in the endodermis called the Casparian strip. At this point, water must again enter the cells by osmosis, allowing plants to filter solutes in the water before it enters the vascular tissue in the root core, or stele.

Adventitious roots form on plants in locations other than the lower axis of the plant. These include the prop roots of corn (*Zea mays*) and roots that form from plant cuttings. Large central roots, like those of carrots (*Daucus carota*), are called taproots. Thin, highly branched root systems like those of grass are fibrous roots.

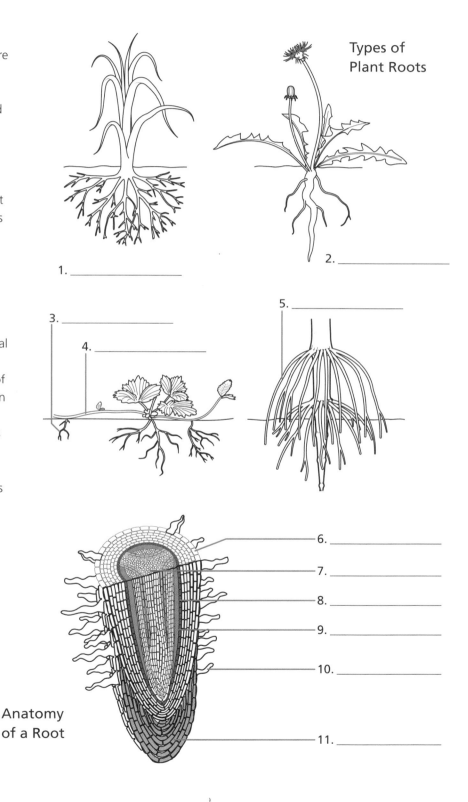

Types of Plant Roots

1. _____

2. _____

3. _____

4. _____

5. _____

Anatomy of a Root

6. _____

7. _____

8. _____

9. _____

10. _____

11. _____

Answers

1. fibrous root system, 2. tap root, 3. adventitious root, 4. stolon, 5. prop roots, 6. epidermis, 7. vascular cylinder, 8. endodermis, 9. pericycle, 10. root hair, 11. root cap

Pressure-Flow Hypothesis

The pressure-flow hypothesis explains how solutes such as sugars move through phloem. Overall, the sugar solution in the phloem moves from areas of high turgor pressure to areas of lower turgor pressure. Plants create these differences in pressure by the transport of sugars from source organs into the phloem and from the phloem into sink organs.

Plant organs such as leaves that carry out photosynthesis are sources of sugar for the entire plant. Cells from these source organs transport sugar in the form of sucrose into the phloem. This raises the solute concentration, leading to the influx of water by osmosis. The movement of water into the sieve elements increases the local turgor pressure in the sieve tubes, which causes the sugar solution to begin to flow away from the high-pressure zone.

As the sugar solution flows through the phloem, plant cells that need sugar can move it out of the phloem and into their cytoplasm. For example, actively growing parts of the plant need sugars for energy, and plants store sugar in fruits and starchy roots. These plant parts are sugar sinks. As they remove sugar from the phloem, the solute concentration decreases. This results in water leaving the phloem and moving either into the tissues that have more solutes or the xylem, which in turn lowers the local turgor pressure in the phloem near the sinks.

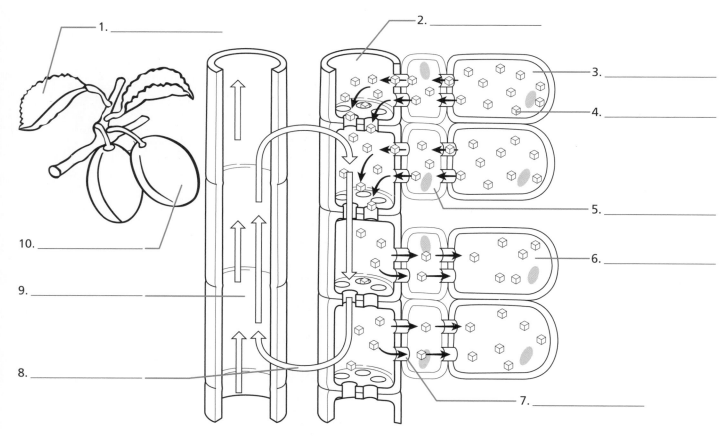

1. _____
2. _____
3. _____
4. _____
5. _____
6. _____
7. _____
8. _____
9. _____
10. _____

Process by which Sugar Is Transported in Plants

Answers

1. leaf (source), 2. phloem, 3. source cell, 4. sucrose, 5. companion cell, 6. sink cell, 7. plasmodesmata, 8. water movement 9. xylem, 10. developing fruit (sink)

Phototropism

Phototropism is the growth of an organism toward or away from light. In plants, it is a response to signals from both light and the plant hormone auxin. This is one of the many examples of how plants use light, not just as energy for photosynthesis but as a source of information about their environment. Plants acquire this information through their photoreceptors, which absorb different types (wavelengths) of light and then trigger important events in plant development. Phototropism is an important plant response because it allows plants to maximize their ability to capture light for photosynthesis.

The photoreceptors that trigger phototropism, called phototropins, absorb blue light. When they detect the presence of light, phototropins affect the distribution of the hormone auxin. Auxin has many effects on plant growth, one of which is to stimulate plant cells to elongate. In shoots that are shaded on one side, phototropins cause the development of an auxin gradient, with the higher concentration of auxin on the shaded side of the stem. The cells on the shaded side elongate more than the cells on the sunny side, causing the stem to curve toward the light, which is called positive phototropism.

Process of Phototropism

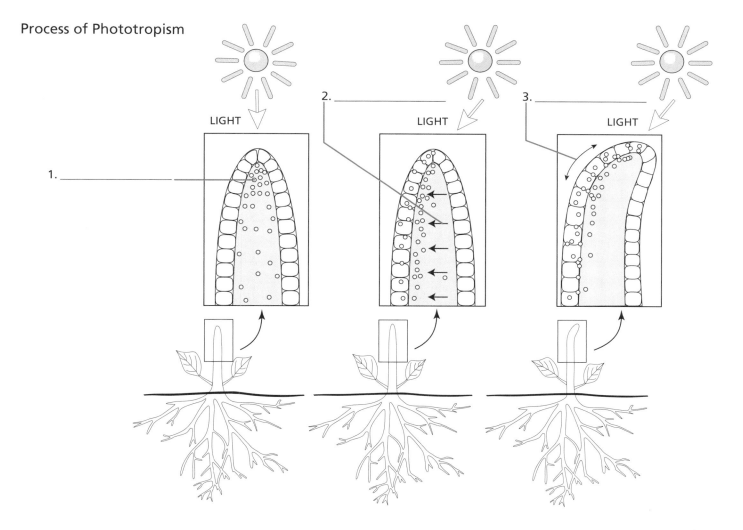

Answers

Phytochrome Responses

Phytochromes are photoreceptors that absorb both red and far-red wavelengths of light. The inactive form of the photoreceptor, called Pr, absorbs red light (~660 nm). The active form, Pfr, absorbs far-red light (~730 nm). When Pr molecules absorb red light, they convert to Pfr. If plants are exposed to far-red light, Pfr converts back to Pr. In darkness, Pfr also converts slowly back to Pr.

Plant Responses to Changes in the Hormone Phytochrome

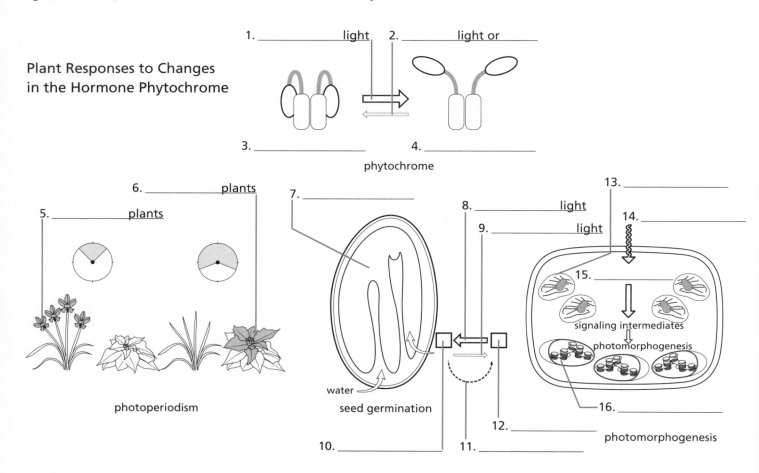

Many plants time their flowering by measuring the photoperiod, or relative amount of day and night. In plants called short-day plants, Pfr inhibits flowering. Therefore, these plants flower only when Pfr levels are very low, which occurs when nights are long and days are short. Long-day plants behave in the opposite manner. In these plants, Pfr activates flowering, so these plants flower when days are long and nights are short.

Phytochromes play a role in light-dependent development, which scientists call photomorphogenesis. Some plants require light for seed germination. When the seeds are exposed to light, Pr converts to Pfr, causing an increase in the plant hormone gibberellin and stimulating cells to turn on the gene for the enzyme α-amylase. This enzyme converts the stored starch into sugars for the growing embryo. Also, plants grown in the dark appear yellow but turn green in the presence of light. Here, the conversion of Pr to Pfr signals cells to turn on genes necessary for the production of chlorophyll, the green pigment in chloroplasts.

Answers

1. red, 2. far-red, 3. dark, 4. inactive Pr, 5. active Pfr, 6. short-day, 7. long-day, 8. red, 9. far-red, 10. Pfr, 11. dark, 12. Pr, 13. proplastid, 14. red light, 15. Pfr, 16. chloroplast

Acid-Growth Hypothesis

The acid-growth hypothesis explains how auxin acts to trigger cell elongation in plants. Auxin stimulates protein pumps called proton-ATPases (H+-ATPase), which use energy from ATP to pump protons across cell membranes. In response to auxin, these pumps actively transport protons across the plasma membrane to the outside of the cell. The increased acidity outside the cell activates pH-dependent proteins called expansins. The expansins loosen the bonds between cellulose molecules in the cell wall, making it more flexible. Water moves into cells by osmosis, resulting in increased turgor pressure. This pressure pushes the cell wall outward, and the cell expands.

How Cell Walls Loosen to Enable Plant Cells to Grow

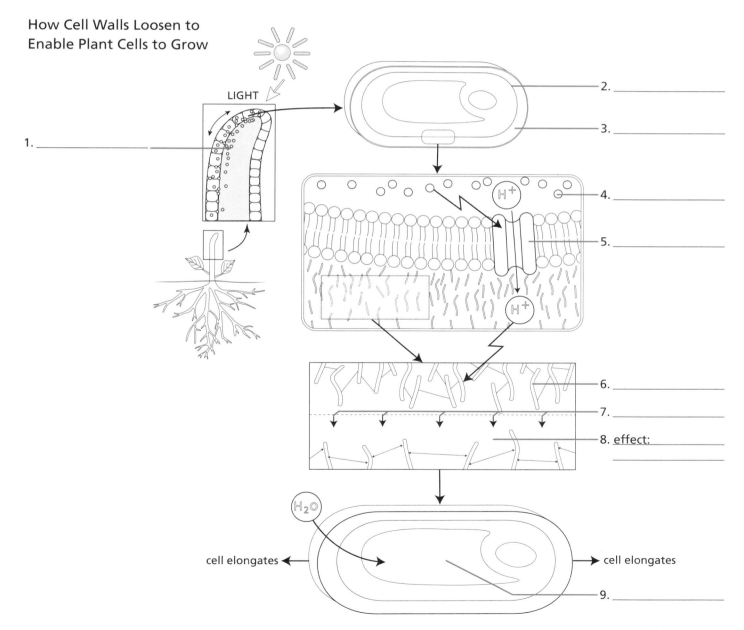

LIGHT

1. _____

2. _____

3. _____

4. _____

5. _____

6. _____

7. _____

8. effect: _____
_____

cell elongates ← → cell elongates

9. _____

Answers

Gravitropism

Gravitropism is the growth of a plant in response to gravity. If a plant is laid down on its side, it will adjust by bending its stem upward and its roots downward. In order for plants to respond to gravity, they must be able to sense it. Plants sense the direction of the gravitational pull through the movement of statoliths, small starch-filled organelles located in specialized cells called statocytes. Gravity pulls the statoliths to one side of the cell, triggering the rearrangement of auxin transporters in the cell membrane. The transporters move auxin into the cell, creating a gradient of auxin with a higher concentration near the statoliths.

   Although both roots and shoots rely on the same mechanism to sense gravity, their responses are opposite. Auxin inhibits the elongation of cells in the root, resulting in less growth adjacent to the statoliths and downward curvature of the root toward the direction of gravity. In shoots, auxin stimulates cell elongation, producing faster growth near statoliths so that stems curve away from gravity. Thus, scientists say that roots show positive gravitropism, while stems show negative gravitropism.

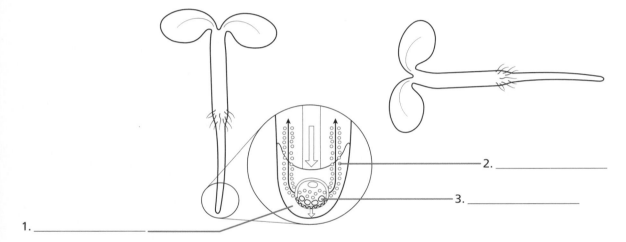

1. _____

2. _____

3. _____

Process by which Plants Respond to Gravity

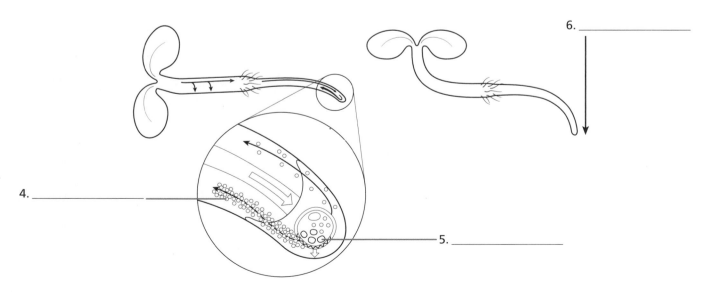

4. _____

5. _____

6. _____

Answers

Seed Germination

Seed germination begins when dry seeds take in water and results in the development of a new plant with a defined shoot and root axis. The uptake of water, called imbibition, occurs in three phases: an initial phase of rapid absorption; a second phase, when absorption plateaus; and a third phase, when uptake is rapid again after germination is complete and the new plant begins to grow.

During the first phase, the molecules in the seed rehydrate, swelling the seed and splitting the seed coat. Levels of the dormancy-inducing hormone abscisic acid fall, and the cells inside the seed break out of dormancy. The cells become metabolically active, using cellular respiration to produce ATP and synthesizing proteins necessary for the early events in seed germination from pre-existing mRNAs. Cells in the embryo produce gibberellin, which turns on the genes for the production of the enzyme α-amylase. This enzyme catalyzes the breakdown of stored starch into sugars that can provide energy and building material for the growing embryo.

During the second phase of imbibition, cells begin producing the materials needed for support of the embryo, such as new mitochondria and proteins needed for growth. Water flows into the cells of the first root, called the radicle, and its cells elongate. The radicle emerges from the seed to complete germination.

Signals Involved in Seed Germination in Barley

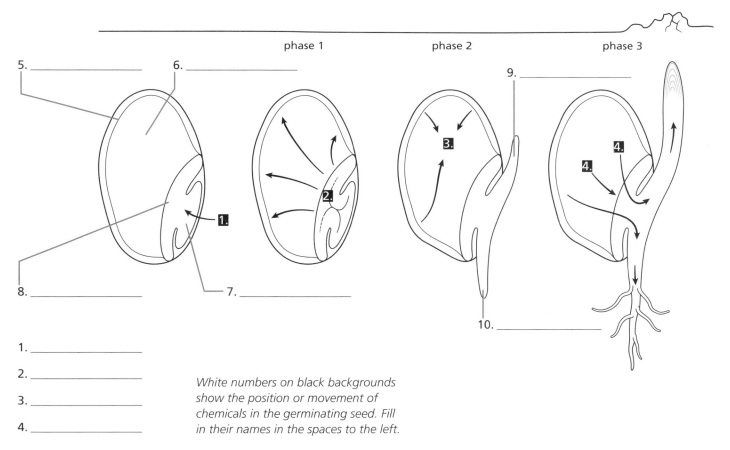

phase 1   phase 2   phase 3

5. _____

6. _____

9. _____

8. _____

7. _____

10. _____

1. _____

2. _____

3. _____

4. _____

White numbers on black backgrounds show the position or movement of chemicals in the germinating seed. Fill in their names in the spaces to the left.

Answers

1. water, 2. gibberellin (gibberellic acid), 3. α-amylase, 4. sugars, 5. aleurone, 6. endosperm, 7. embryo, 8. cotyledon (scutellum), 9. plumule, 10. radicle

Alternation of Generations

Plant life cycles appear quite different from those of animals. In animals, diploid adults produce microscopic haploid gametes that must fuse to create a new diploid individual. In plants, however, the diploid and haploid generations can be more or less independent from each other depending on the species. If the life cycle of some plants were transposed onto a human, for example, it would be as if the human gamete (egg or sperm) went off and had a life of its own before settling down to fertilization and the creation of a new diploid organism.

The diploid generation in plants is the sporophyte. Sporophytes have two complete sets of chromosomes, so they divide by meiosis to produce haploid spores, which have just one set of chromosomes. These spores grow by mitosis to produce the gametophyte generation. Gametophytes produce gametes, but because gametophytes are already haploid, they produce these gametes by mitosis. When two gametes fuse together, they form a diploid zygote, which divides by mitosis to grow into a new sporophyte. Because plants alternate between two forms, the sporophyte and the gametophyte, scientists describe their life cycle as alternation of generations. In some plants, the sporophyte generation is larger and more visible, while in others, the gametophyte is the dominant form.

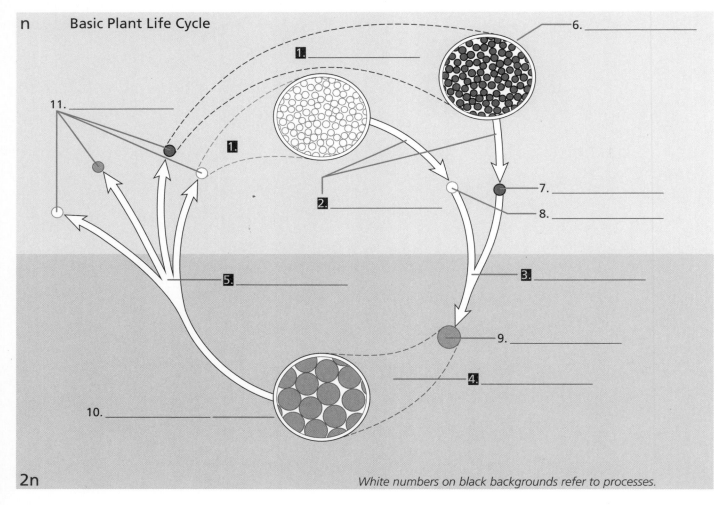

Basic Plant Life Cycle

n

6. _____

1. _____

11. _____

1. _____

2. _____

7. _____

8. _____

5. _____

3. _____

9. _____

4. _____

10. _____

2n

White numbers on black backgrounds refer to processes.

Answers

1. mitosis, 2. mitosis, 3. fertilization, 4. mitosis, 5. meiosis, 6. gametophyte, 7. male gamete, 8. female gamete, 9. zygote, 10. sporophyte, 11. spores

Moss Life Cycle

In the moss life cycle, the gametophyte generation is dominant. The spongy green plants growing on trees and the forest floor are haploid. In order to reproduce sexually, mosses form special male and female structures at the top of their stalks. Mosses produce eggs inside of vase-shaped chambers called archegonia and produce sperm inside of club-shaped structures called antheridia. Some mosses are monoecious, meaning they produce both male and female parts, while others are dioecious, meaning that some plants are entirely female, while others are entirely male.

Mosses require water for sexual reproduction because their sperm need to swim to find the eggs. Fertilization produces a zygote inside the archegonia, which then divides by mitosis to produce a slender, stalked sporophyte. Part of the gametophyte remains on top of the sporophyte like a small hat called the calyptra. The tip of the sporophyte swells to form the capsule; the stalk is called the seta. Meiosis occurs inside the capsule, producing many haploid spores. When the spores are mature, the operculum, or lid, of the capsule blows off to release the spores to the wind. A ring of teeth-like tissue around the edge of the capsule, called the peristome, flexes with changes in humidity to help disperse the spores. Once the spores land, they grow by mitosis to produce a new gametophyte.

Life Cycle of a Moss

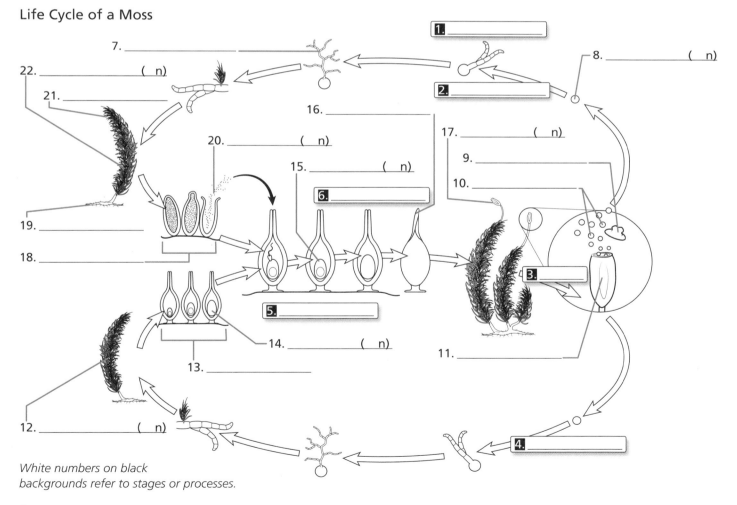

White numbers on black backgrounds refer to stages or processes.

Answers

1. germination, 2. mitosis, 3. meiosis, 4. mitosis, 5. fertilization, 6. mitosis, 7. protonema, 8. spore, 1, 9. operculum, 10. spores, 11. capsule, 12. female gametophyte, 1, 13. archegonia, 14. egg, 1, 15. zygote, 2, 16. calyptra, 17. sporophyte, 2, 18. antheridia, 19. rhizoids, 20. sperm, 1, 21. leaves, 22. male gametophyte, 1

Fern Life Cycle

In ferns, sporophytes dominate the landscape, and gametophytes, which are only about as big as a human fingernail, are easily missed. Fern leaves, called fronds, grow from buds on sprawling stems called rhizomes. Young fronds appear curled up like the ornamental end of a violin, which is why they're often called fiddleheads. Spore-producing structures called sporangia form on the underside of mature fronds in small clusters called sori. Some sori have a protective cover called an indusium. Each sporangium produces many haploid spores by meiosis. When the spores are mature, the sporangia dry out until tension tears them open and frees the spores.

Fern spores germinate in moist soil to produce small heart-shaped gametophytes, which are anchored to the soil by rhizoids. Archegonia develop near the notch at the top of the gametophyte, and antheridia develop near the rhizoids at the base. Mitosis produces an egg in the archegonia and many sperm in the antheridia. Sperm swim to find the eggs, fertilizing them to produce a new zygote. The zygote divides by mitosis to grow a new sporophyte frond.

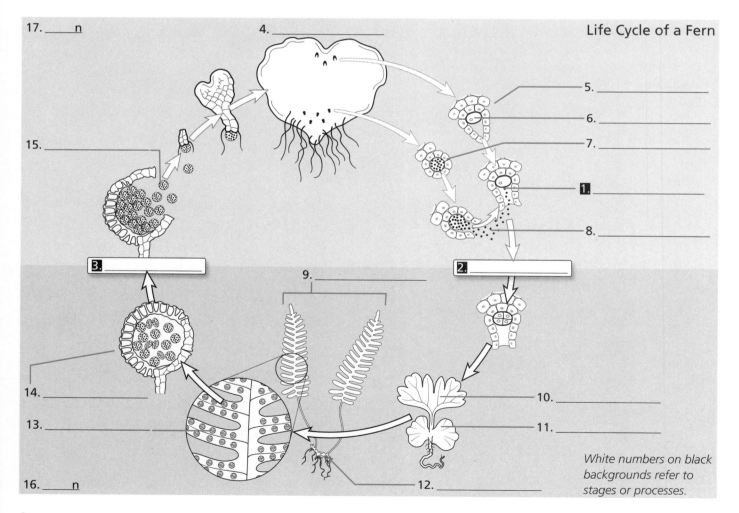

Life Cycle of a Fern

17. _____ n

4. _____

5. _____

6. _____

7. _____

1. _____

8. _____

15. _____

3. _____

9. _____

2. _____

14. _____

13. _____

10. _____

11. _____

16. _____ n

12. _____

White numbers on black backgrounds refer to stages or processes.

Answers

Gymnosperm Life Cycle

The gymnosperms that you see in the landscape, such as pines, cedars, and junipers, are all sporophytes. In gymnosperms, gametophytes consist of only a few cells. In coniferous gymnosperms, gametophytes form inside cones (strobili). Gymnosperms are heterosporous, meaning they produce male and female spores on separate structures. Some species are dioecious, while others are monoecious.

The cones of gymnosperms are diploid. In the female or ovulate cones, meiosis occurs inside the megasporangium to produce the megaspore and three other cells, which degenerate. The megaspore divides by mitosis to produce a microscopic female gametophyte. In male or staminate cones, meiosis occurs in microsporangia to produce microspores. The microspores divide by mitosis to produce the male gametophytes, which are the pollen grains.

Mitosis occurs inside gametophytes to produce gametes. The female gametophyte produces eggs inside archegonia. Mitosis inside the pollen grains produces sperm. The microsporangia split open, releasing pollen to the wind. When pollen lands on a female gametophyte, a pollen tube grows down through the archegonium to reach the ovule. The sperm travel down the pollen tube and fertilize the ovule, forming a zygote that divides by mitosis to produce an embryo. Seed development follows, and the ovule wall hardens to become a protective seed coat. When the seed reaches an appropriate environment, it germinates and the embryo grows by mitosis into a new sporophyte.

Life Cycle of a Gymnosperm

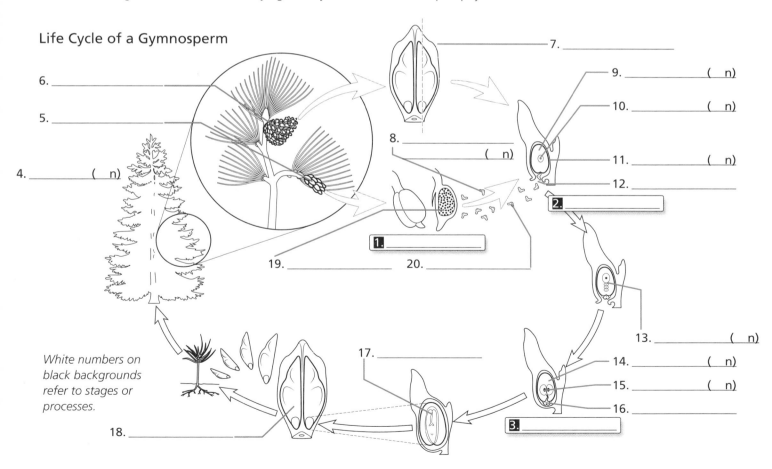

White numbers on black backgrounds refer to stages or processes.

6. _____

5. _____

4. _____ (n)

7. _____

9. _____ (n)

10. _____ (n)

11. _____ (n)

12. _____

8. _____ (n)

2. _____

1. _____

19. _____   20. _____

13. _____ (n)

14. _____ (n)

15. _____ (n)

16. _____

3. _____

17. _____

18. _____

Answers

Flower Structure

Flowers are the reproductive structures of angiosperms. The best way to identify the parts of a flower is to start at the bottom and identify each whorl of structures as you move inward. The place at the bottom of the flower where all the structures meet is the receptacle, and the stalk of the flower is the pedicel. As you move inward, the first whorl of structures, called the calyx, contains the leaf-like sepals. The next whorl, called the corolla, contains the petals. These are also leaf-like but are usually brightly colored to attract pollinators.

    The remainder of the flower contains the male and female parts. The first whorl inside the petals contains the male structures, called stamens, which consist of a thin stalk called a filament and pouch-like structures called anthers. Pollen forms inside the anthers. The female parts are in the center of the flower. Each separate female part is called a carpel. Some flowers have a whorl of many separate carpels, while in others the carpels fuse together to form one central female structure called a pistil. The swollen base of the pistil is the ovary, where ovules develop. Rising up from the ovary is a stalk called the style, the tip of which is called the stigma.

Basic Parts of a Flower

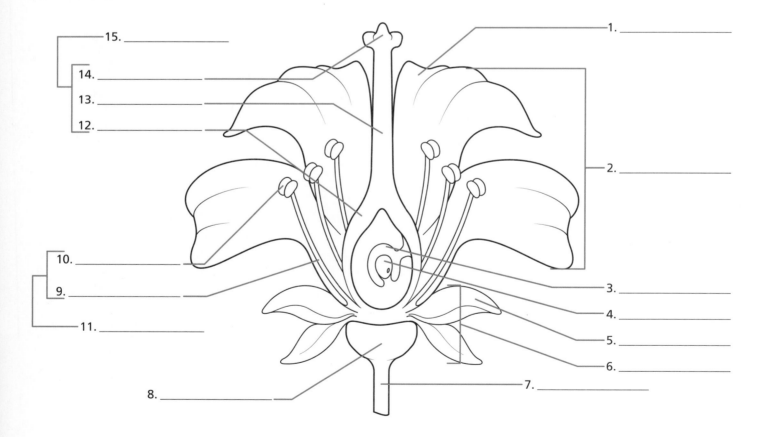

15. _____

14. _____

13. _____

12. _____

10. _____

9. _____

11. _____

8. _____

1. _____

2. _____

3. _____

4. _____

5. _____

6. _____

7. _____

Answers

Angiosperm Life Cycle

Like gymnosperms, sporophytes dominate the angiosperm life cycle and gametophytes are microscopic. Sporophytes produce flowers, which can have both male and female parts. Meiosis occurs inside microsporangia in the anther to produce haploid microspores. The microspores divide by mitosis to form the male gametophyte, which is called pollen. In the ovary of the flower, megasporangia form inside the ovules. Meiosis inside the megasporangia produce four megaspores, three of which degenerate. The remaining megaspore divides by mitosis to produce a female gametophyte made of seven cells: one egg with two companion cells called synergids, one central cell with two polar nuclei, and three antipodal cells.

Angiosperms have a unique type of fertilization called double fertilization. When pollen lands on the stigma of the flower, pollen tubes grow down through the style so that sperm can reach the female gametophytes inside the ovule. Each pollen grain delivers two sperm to the ovule. One sperm fuses with the egg, while the other fuses with the two polar nuclei to form a triploid tissue called endosperm. The zygote divides by mitosis to produce the embryo. The ovule develops into a seed, and the ovule wall becomes a protective seed coat. The ovary develops into a fruit that encloses the seeds.

Life Cycle of an Angiosperm

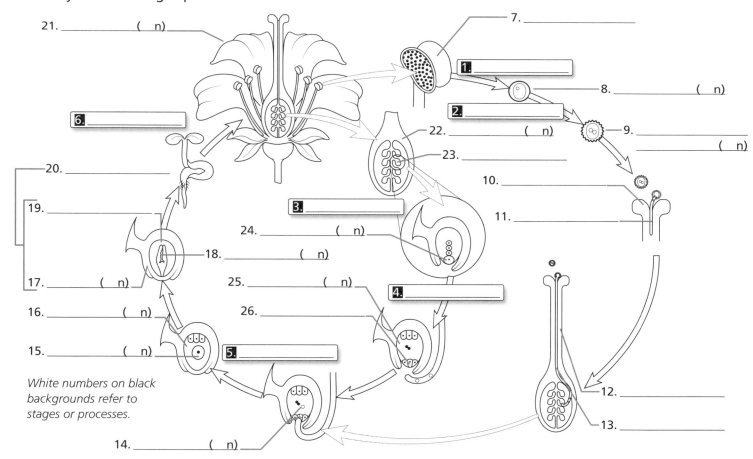

White numbers on black backgrounds refer to stages or processes.

Answers

Fruit Development

Anything that makes a flower produces a fruit. Some fruits are soft and fleshy, such as plums; others are dry and hard, such as walnuts; and some are light and airy, such as the parachutes of dandelions. The purpose of fruit is to protect and distribute seeds. Fleshy fruits typically attract animals that eat the fruits, while dandelion parachutes fly away on a passing breeze.

After fertilization, the ovary wall develops into the fruit wall, or pericarp. In true fruits, only the ovary wall becomes pericarp, but in accessory fruits, other flower parts may be incorporated. In apples, for example, the inner core forms from the ovary, while the fleshy outer part of the apple develops from tissue at the top of the receptacle. The pericarp can have up to three layers: the outermost epicarp (exocarp), the middle mesocarp, and the inner endocarp. In fleshy fruits, the epicarp is the skin of the fruit and the mesocarp is the fleshy portion. In some fruits, the endocarp forms a hard layer around the seeds.

Simple fruits form from a single ovary or pistil. In some fruits, such as raspberries, separate carpels in a flower fuse during fruit development to form an aggregate fruit. In other plants, such as pineapple, the carpels from an entire stalk of flowers may fuse to form a multiple fruit.

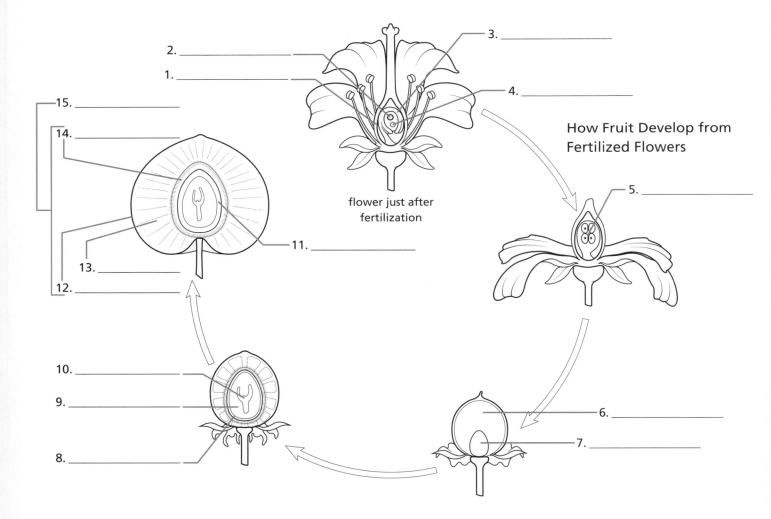

2. _____

1. _____

3. _____

4. _____

flower just after fertilization

How Fruit Develop from Fertilized Flowers

5. _____

11. _____

15. _____

14. _____

13. _____

12. _____

10. _____

9. _____

8. _____

6. _____

7. _____

Answers

Bilateral vs. Radial Symmetry

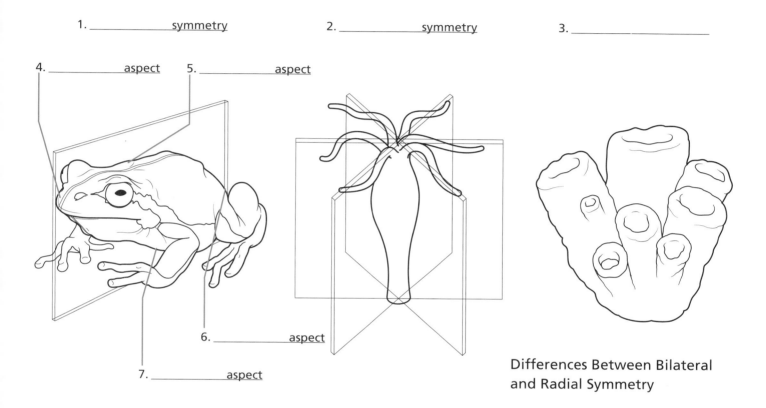

1. _____ symmetry

2. _____ symmetry

3. _____

4. _____ aspect

5. _____ aspect

6. _____ aspect

7. _____ aspect

Differences Between Bilateral and Radial Symmetry

Most animals exhibit body symmetry such that their bodies can be divided into nearly identical halves around at least one plane. The more ancient lineages of animals, such as the cnidarians, ctenophores, and some sponges, exhibit radial symmetry; their bodies can be divided into symmetrical pieces around two or more planes. If you imagine cutting a sea anemone vertically into quarters like you would a pie, for example, you'd get four very similar pieces. Echinoderms also exhibit radial symmetry, but scientists believe it evolved independently in the group.

Bilateral symmetry evolved later than radial symmetry and is found in almost all groups of protostomes and deuterostomes. In organisms with bilateral symmetry, called bilaterians and including humans, you find only one plane of symmetry, most often running vertically from the anterior to the posterior end. Organisms with bilateral symmetry often show cephalization, which is the development of a head or anterior region with concentrated sensory receptors and feeding structures. The mass of neurons that process sensory information is the cerebral ganglion or brain.

Answers

1. bilateral, 2. radial, 3. asymmetry, 4. anterior, 5. dorsal, 6. posterior, 7. ventral

Development of Body Cavities

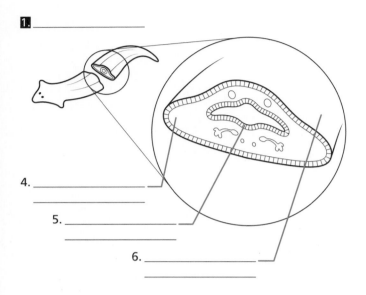

1. _____

4. _____
_____

5. _____
_____

6. _____
_____

The body plan of bilaterian animals is commonly referred to as a tube within a tube; the inner tube is the gut of the animal, which forms from endoderm and has a mouth at one end and an anus at the other, and the outer tube is the external covering of the animal, which forms from ectoderm and consists of the skin and nervous tissue. In between these two tubes is the mesoderm, which forms the muscles and organs.

In addition to mesoderm, some bilaterians develop a fluid-filled body cavity in between the outer and inner tubes of their bodies. If this cavity is completely lined with mesoderm, scientists call it a coelom, and if it's only partially lined with mesoderm, it's a pseudocoelom. Animals that don't develop a fluid-filled body cavity are said to be acoelomate.

Coeloms and pseudocoeloms provide a space for internal organs, allowing them to move independently from the rest of the body. This fluid-filled cavity also allows for the circulation of gases and nutrients and provides hydrostatic pressure.

Label body regions and their embryonic tissue type. White numbers on black backgrounds are body types.

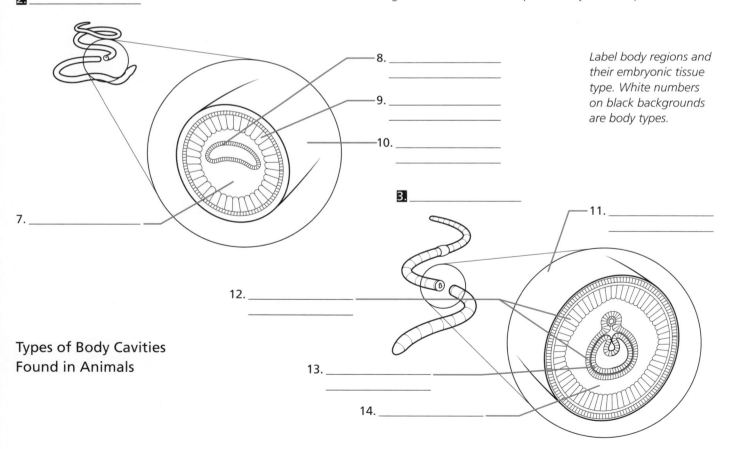

2. _____

8. _____
_____

9. _____
_____

10. _____
_____

7. _____

3. _____

11. _____
_____

12. _____
_____

13. _____
_____

14. _____

Types of Body Cavities Found in Animals

Metamorphosis

Metamorphosis refers to abrupt and dramatic changes that occur during the development of some animals, such as insects, amphibians, and sea urchins. The various stages of the animal may occupy different habitats and show different feeding strategies and behavior. In these animals, zygotes develop into initial stages called larvae. Larvae grow and transform into juveniles, which may look like adults but are sexually immature. After further growth, juveniles mature into adults, which are capable of sexual reproduction.

In some animals, metamorphic transformations involve ecdysis, or molting, to remove the exoskeleton or skin. In insects that undergo incomplete metamorphosis (hemimetabolous), each developmental stage is called an instar, and immature instars are called nymphs. In insects that undergo complete metamorphosis (homometabolous), immature stages are called larvae. Larvae enter an inactive phase inside a pupa (called a chrysalis in butterflies) and then emerge as adults.

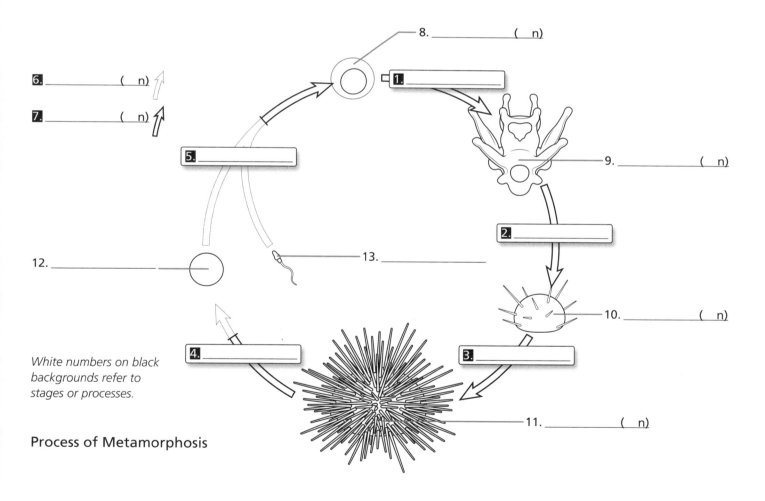

White numbers on black backgrounds refer to stages or processes.

Process of Metamorphosis

Chordate Characteristics

Chordates are animals that show four key characteristics at some point in their life cycles. Pharyngeal gill slits are openings or pouches in the throat that develop into gill arches in bony fish and into the jaw and inner ear in terrestrial animals. The dorsal hollow nerve cord is nervous tissue that forms a tube along the dorsal side of the animal. The notochord is a flexible but firm rod that runs the length of the body. A muscular post-anal tail is a tail that extends past the anus and contains muscle.

The invertebrate chordates are the cephalochordates, or lancelets, and the urochordates, or tunicates. Cephalochordates retain all four chordate characteristics into adulthood, but some urochordates, such as the sea squirts, retain only the pharyngeal gill slits.

The vertebrate chordates are the cartilaginous fish, bony fish, amphibians, mammals, reptiles, and birds. In these animals, the dorsal hollow nerve cord is retained into adulthood as the spinal cord. Pharyngeal gill slits and the muscular post-anal tail are present in all embryos but are lost during the development of many species, including humans. Notochords are also present in all embryos, and although they don't persist into the adult stage of most vertebrates, they help trigger the development of the vertebrae that protect the spinal cord.

Four Key Characteristics of Chordates

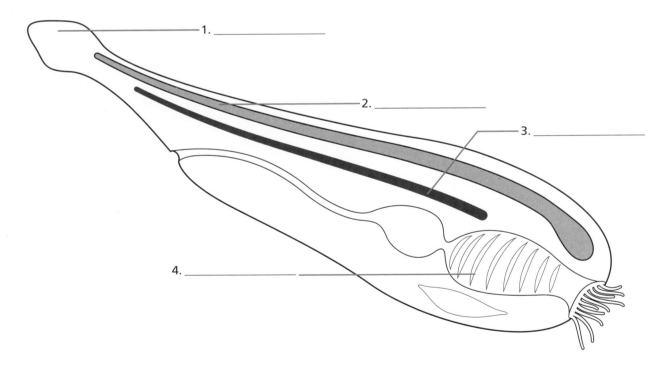

1. _____

2. _____

3. _____

4. _____

Answers

1. post-anal tail; 2. dorsal hollow nerve cord; 3. notochord; 4. pharyngeal (gill) slits

Amniotic Egg vs. Placenta

Amniotic eggs and placentas support the growth and development of the embryo in dry, terrestrial environments. They facilitate the exchange of gases and the flow of nutrients to the embryo while removing wastes. The amniotic egg evolved first in the common ancestor of reptiles and mammals. Amniotic eggs have an external protective covering that protects against drying and four internal membranes that provide cushioning and surface area for gas exchange. The amnion surrounds the embryo, the allantois contains the waste materials, the yolk sac houses the nutrients, and the chorion encloses all of the other membranes.

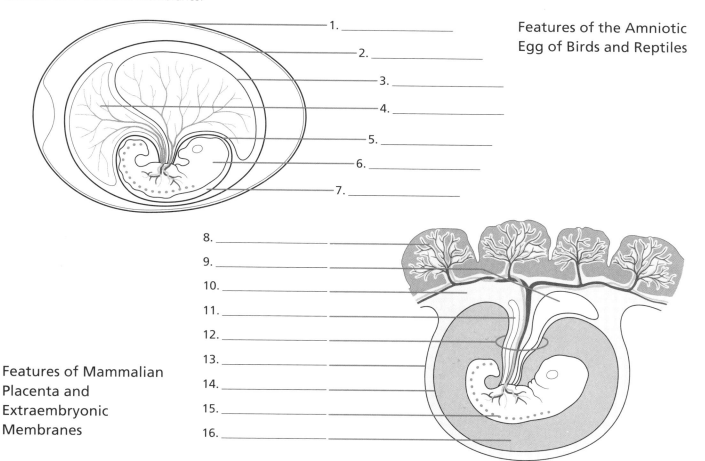

Features of the Amniotic Egg of Birds and Reptiles

1. _____
2. _____
3. _____
4. _____
5. _____
6. _____
7. _____

Features of Mammalian Placenta and Extraembryonic Membranes

8. _____
9. _____
10. _____
11. _____
12. _____
13. _____
14. _____
15. _____
16. _____

The placenta is unique to a large and diverse group of animals called placental mammals. These mammals produce a placenta in the uterus or oviduct during pregnancy. The placenta is a combination of maternal and embryonic tissue that forms from the same membranes as those found in the amniotic egg. The amnion forms the inner layer of the placenta and contains amniotic fluid, which surrounds and protects the fetus. The chorion forms the outer layer of the fetal portion of the placenta, surrounding the amnion and other membranes. It transfers nutrients from the mother to the fetus. The allantois and yolk sac develop into the umbilical cord, which connects the fetus to the placenta. Remnants of these membranes remain in the cord.

Answers

1. shell, 2. chorion, 3. yolk sac, 4. allantois, 5. amnion, 6. embryo, 7. amniotic cavity, 8. placenta (maternal portion), 9. yolk sac, 10. placenta (fetal portion), 11. allantois, 12. umbilical cord, 13. chorion, 14. amnion, 15. embryo, 16. amniotic cavity

Homeostasis

Homeostasis is the maintenance of internal chemical and physical conditions in a range that supports life even in the face of changing environmental conditions. This ability is critical because extreme internal conditions could denature proteins, disrupting metabolism and other essential functions. To maintain homeostasis, organisms have systems for sensing and responding to changes in conditions such as temperature, pH, and chemical concentrations.

   The regulatory systems that maintain homeostasis have three components: a sensor, a control center, and an effector. Sensors are receptors that detect changes in the external or internal environment. They relay this information to a control center that processes the information and compares it to desired conditions. If a response is required, the control center sends a signal to an effector that will trigger a change in the organism. To maintain homeostasis, effectors often resist or reverse the original change. Because the change in conditions triggered a response to counteract this change, scientists call this type of response negative feedback.

   Scientists often compare biological homeostasis to the maintenance of room temperature by a system that uses a thermostat. The thermostat is set for a desired condition, and when the temperature gets too high or too low, a sensor detects the change and triggers activation of heating or cooling to return the room to the set point.

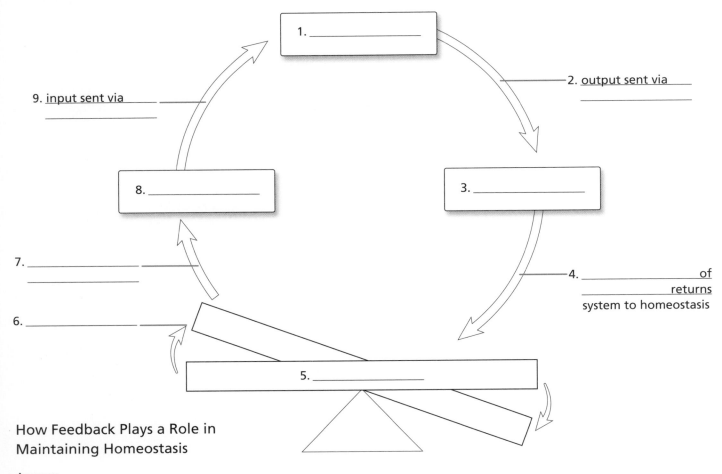

1. _____

2. output sent via _____
   _____

9. input sent via _____
   _____

8. _____

3. _____

7. _____
   _____

4. _____ of
   _____ returns
   system to homeostasis

6. _____ _____

5. _____

How Feedback Plays a Role in Maintaining Homeostasis

Answers

Gas Exchange

Animals exchange gases with their environment in order to acquire the oxygen they need for cellular respiration and release carbon dioxide as waste. In order for gas exchange to occur, gas is drawn in from the environment and passed across a respiratory surface, such as the membranes of gills or lungs. Gases diffuse across this surface according to their concentration gradient and are then circulated around the body via the circulatory system. Oxygen is delivered to cells so that it can be used by mitochondria for cellular respiration. At the same time, carbon dioxide produced as waste by cellular respiration is collected by the circulatory system and delivered back to the respiratory surface so it can be expelled from the body.

Animals that live in wet environments may diffuse gases directly across their external surface membranes, while animals that live in dry environments have specialized organs for respiration. Gills are projections of the body surface or throat that provide a large surface area for gas exchange. Lungs, which occur in most terrestrial vertebrates, are internal organs that allow animals to draw in air and expel it. Gas exchange occurs across the moist surface of the lung tissue. The tracheae of insects are branching, air-filled tubes that pass very close to cells, enabling gas exchange directly across cell membranes.

Main Steps in Gas Exchange and Examples of Respiratory Systems in Animals

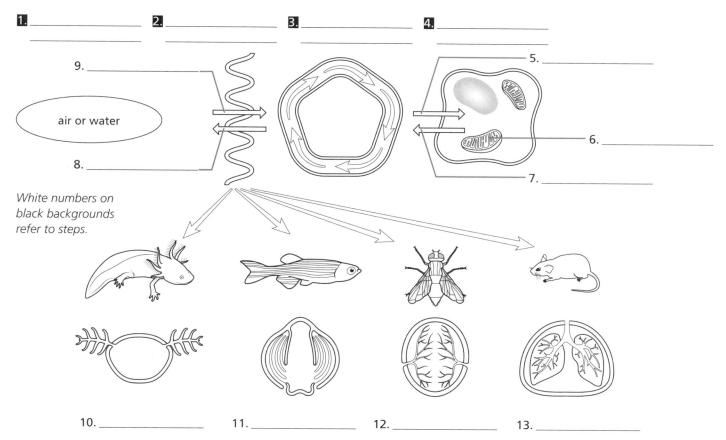

1. _____
2. _____
3. _____
4. _____
5. _____
6. _____
7. _____
8. _____
9. _____

air or water

White numbers on black backgrounds refer to steps.

10. _____
11. _____
12. _____
13. _____

Answers

Close-Up of Countercurrent Exchange

Mechanism of Gas Exchange by Countercurrent Exchange

Countercurrent exchange can occur when two fluids flow past each other in opposite directions. In fish gills, the mechanism greatly increases the efficiency of gas exchange. Water flows in one direction through the gills, passing across each lamella in the direction opposite that of blood flow through the capillaries. As water flows across the blood vessel, oxygen diffuses out of the water and into the blood. Because the two solutions flow in opposite directions, the concentration of oxygen is always higher in the water than in the blood, allowing diffusion to occur across the entire gill surface. This maximizes the amount of oxygen that can be transferred by diffusion from the water to the blood.

4. _____

3. _____

2. _____

1. _____

5. _____

6. _____ -poor blood

lamella

13. _____ -rich blood

7. _____

Water flows in through the mouth of the fish and across the gills. The water flows across each lamella from the side with the highest concentration of oxygen in the blood to the side with the lowest concentration of oxygen in the blood. Although the concentration of oxygen in the water decreases as it flows, the water is moving toward blood with a lower concentration of oxygen, so the water still has the higher amount. If the two solutions flowed in the same direction, they would reach a point where their oxygen concentrations were almost equal and little diffusion would occur.

8. _____

_____

12. _____

9. _____

10. _____

11. _____

countercurrent exchange

Labels 10–13 describe levels of oxygen in water and blood.

Answers

Osmoregulation

Osmoregulation is the process organisms use to control the relative amounts of water and solutes in their bodies. Some marine invertebrates, called osmoconformers, have an internal solute concentration that's nearly identical to seawater. Because they're isotonic to their surroundings, these animals don't need to adjust their water and solute concentrations. Bony fish and terrestrial animals, however, maintain an internal solute concentration that's very different from that of their environment. For these animals, osmoregulation is an important part of homeostasis.

Marine and freshwater fish face opposite challenges. For marine fish, seawater is much saltier than their body fluids, causing water to leave their bodies by osmosis. They replace this water by drinking large amounts of seawater and getting rid of the excess solutes. They also conserve water by producing very little urine. In contrast, freshwater fish gain water by osmosis because their body fluids have a higher solute concentration than their environment. To balance this influx, these fish don't drink, produce large amounts of dilute urine, and actively transport ions into their bodies.

Terrestrial animals constantly lose water to their environment by evaporation. Most replace this water with the water obtained through food and drink. Desert animals, however, face special challenges due to the scarcity of water. To counteract this, some produce very small, highly concentrated volumes of urine and other waste.

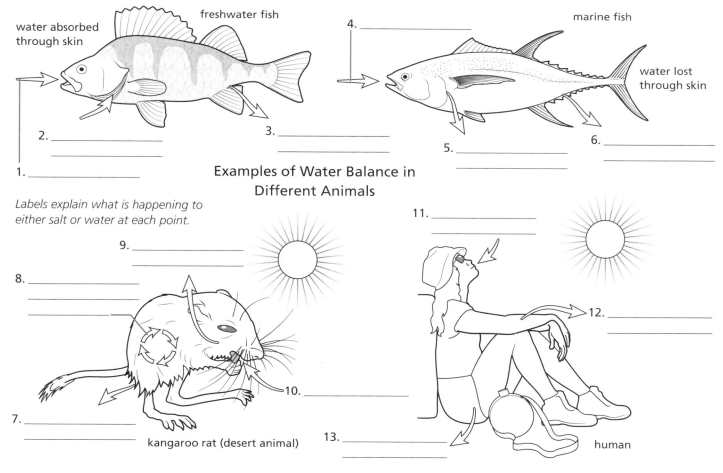

freshwater fish

water absorbed through skin

marine fish

water lost through skin

4. _____

2. _____

3. _____

5. _____

6. _____

1. _____

Examples of Water Balance in Different Animals

Labels explain what is happening to either salt or water at each point.

9. _____

8. _____

11. _____

7. _____

10. _____

12. _____

13. _____

kangaroo rat (desert animal)

human

Answers

1. salt in food; does not drink, 2. salt absorbed by gills and transported to tissues, 3. large volume of dilute urine, 4. drinks seawater, 5. salt removed from blood and excreted by gills, 6. small volume of urine, 7. small volume of urine, 8. most water derived from metabolism of food, 9. does not drink, 10. does not sweat, few sweat glands, 11. water ingested through drink and food, 12. loses water through evaporation, 13. water lost through feces and urine

Circulatory Systems

Circulatory systems move liquid tissues such as blood around the bodies of animals in order to distribute materials such as gases, food, and wastes. Circulatory systems may be open or closed. In both types, a heart pumps the fluid through vessels. In open circulatory systems, however, the fluid is free to leave vessels and enter body cavities, where it touches cells directly. In closed circulatory systems, the fluid remains within vessels the entire time. As a result, open circulatory systems have lower fluid pressures that typically result in slower flow rates. Animals with open circulatory systems also have less control over the direction of flow. When the circulating fluid, called hemolymph, comes into contact with cells, materials can move directly between the two across the cell membrane.

The higher fluid pressure in closed circulatory systems results in faster flow rates that are more suitable for active animals such as vertebrates and burrowing invertebrates. Several types of vessels can be found in these systems. Arteries carry oxygenated blood away from the heart to the body. Veins carry deoxygenated blood back to the heart. Capillaries are very thin vessels that pass close to tissues so that materials can be exchanged. In closed circulatory systems, materials must be exchanged across vessel walls, so the walls of capillaries are only one cell thick.

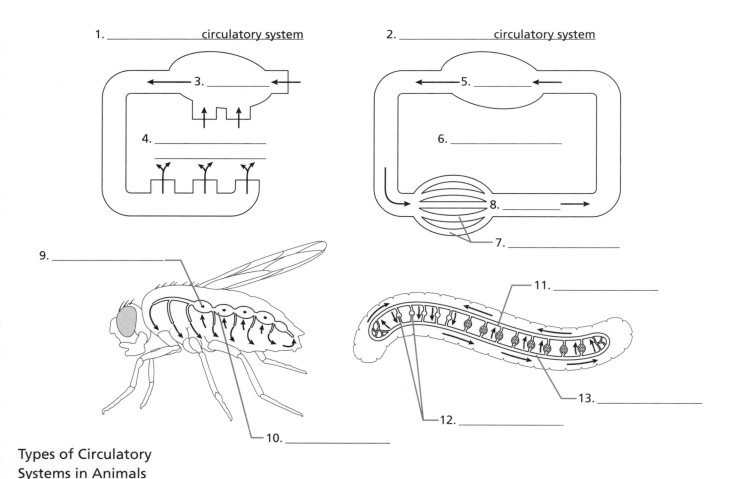

1. _____ circulatory system

2. _____ circulatory system

3. _____
4. _____
5. _____
6. _____
7. _____
8. _____
9. _____
10. _____
11. _____
12. _____
13. _____

Types of Circulatory Systems in Animals

Answers

Nervous Systems

The function of a nervous system is to receive, interpret, and send signals throughout an animal's body. The signals move through the system as electrical impulses conducted by cells called neurons. The organization of neurons in an animal's body depends on the environment and lifestyle of the animal.

Animals that don't have a defined head or tail, such as sea jellies, have web-like arrangements of neurons called nerve nets. Other animals with radial symmetry, such as sea stars, have relatively simple nervous systems. In animals that show cephalization, or a defined head, nervous systems show organization into a central nervous system. Groups of neurons are clustered into bundles called ganglia, the largest of which is the cerebral ganglia, or brain, located at the anterior end or in the head of the organism.

Sensory cells throughout an animal's body collect information. These cells may pass the information onto sensory neurons, or they may be sensory neurons themselves. Sensory neurons relay the information to interneurons in the central nervous system, which integrate the information. The interneurons may pass the signal onto motor neurons, which connect to muscles or glands. The muscles or glands can be stimulated to produce a response to the original sensory signal.

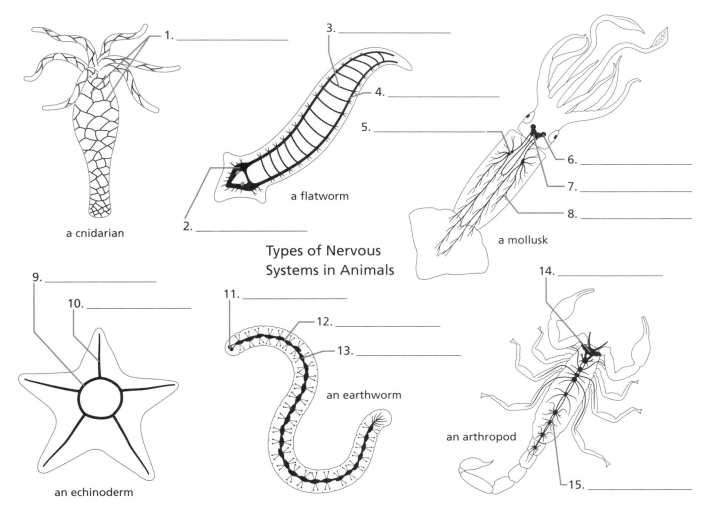

1. _____

a cnidarian

3. _____
4. _____
5. _____
2. _____

a flatworm

6. _____
7. _____
8. _____

a mollusk

Types of Nervous Systems in Animals

9. _____
10. _____

an echinoderm

11. _____
12. _____
13. _____

an earthworm

14. _____
15. _____

an arthropod

Answers

1. nerve net, 2. brain, 3. transverse nerve, 4. nerve trunk, 5. stellate ganglion, 6. optical ganglion, 7. brain, 8. mantle nerves, 9. neural ring, 10. radial nerve, 11. brain, 12. ventral nerve cord, 13. segmental ganglion, 14. brain, 15. ventral nerve cords

Skeletal Systems

Skeletons provide protection and structure to the bodies of animals. They are necessary for movement because they provide a framework for muscles to pull against, enabling animals to change their shape and exert forces. Some animals, like sea jellies and earthworms, have a hydrostatic skeleton that provides resistance based on internal fluid pressure against a flexible body wall. For example, when local muscles in an earthworm contract, they squeeze on the internal fluid, resulting in the extension of another part of the worm.

Endoskeletons occur completely inside the bodies of animals. The bones of these skeletons consist of cells hardened by an extracellular matrix primarily made of calcium phosphate. The place where two bones meet is called a joint. Ligaments are strong, flexible bands of connective tissue such as collagen that connect bones to bones and stabilize joints. Muscles attach to bones in pairs of muscle groups called flexors and extensors. The action of each pair is antagonistic, or opposite, to each other: flexors decrease the angle at a joint, while extensors increase the angle.

Animals such as insects and crustaceans have exoskeletons, which are on the outside of their bodies. The exoskeletons of insects are hardened by the nitrogen-containing polysaccharide chitin, while those of crustaceans contain calcium carbonate. Antagonistic pairs of muscle groups attach to the exoskeleton to enable movement around joints.

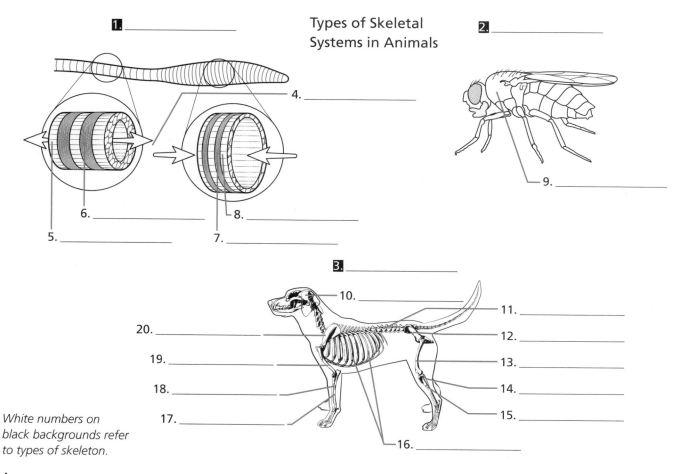

Types of Skeletal Systems in Animals

1. _____

2. _____

4. _____

5. _____

6. _____

7. _____

8. _____

9. _____

3. _____

10. _____

11. _____

12. _____

13. _____

14. _____

15. _____

16. _____

17. _____

18. _____

19. _____

20. _____

White numbers on black backgrounds refer to types of skeleton.

Answers

Digestive Systems

Digestive systems break food down into pieces small enough to be absorbed by cells and then eliminate the wastes. Animals with incomplete digestive systems, such as sea jellies, take in food and eliminate wastes through the same opening, their mouth. Food passes from the mouth into the gastrovascular cavity, where it is digested and cells absorb food molecules. Wastes are then expelled through the mouth. This type of digestive system can be inefficient because food molecules may be excreted along with wastes.

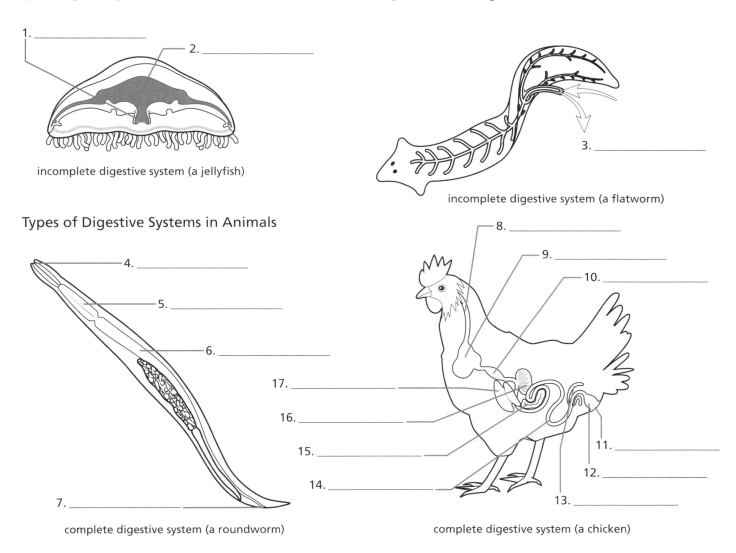

1. _____

2. _____

incomplete digestive system (a jellyfish)

3. _____

incomplete digestive system (a flatworm)

Types of Digestive Systems in Animals

4. _____

5. _____

6. _____

7. _____

complete digestive system (a roundworm)

8. _____

9. _____

10. _____

17. _____

16. _____

15. _____

14. _____

11. _____

12. _____

13. _____

complete digestive system (a chicken)

Most animals have complete digestive systems with two distinct openings, a mouth and an anus. Food is taken in through the mouth and wastes are expelled through the anus. In complete digestive systems, different processes can be localized to separate parts of the system. For example, digestion can occur in one section, while absorption occurs in another. This organization allows for the specialization of structures and enzymes in each area. Ingestion of food and elimination of waste occur at opposite ends of the system, so both can take place independently.

Answers

1. mouth/anus, 2. stomach, 3. food and waste, 4. mouth, 5. pharynx, 6. intestine, 7. anus, 8. esophagus, 9. crop, 10. proventriculus, 11. large intestine, 12. cloaca, 13. caecum, 14. small intestine, 15. pancreas, 16. gizzard, 17. liver

Gametogenesis

Gametogenesis refers to the sequence of cell divisions that produce haploid gametes from diploid cells. In most animals, gametogenesis occurs in special sex organs called gonads. Male gonads are called testes; female gonads are ovaries. Cells in testes undergo spermatogenesis to produce male gametes, or sperm. Cells in ovaries undergo oogenesis to produce female gametes, or ova.

During spermatogenesis, diploid cells in the testes called spermatogonia divide by mitosis. Some of the resulting cells differentiate into primary spermatocytes, which divide by meiosis. Meiosis I produces two haploid cells called secondary spermatocytes; after these cells finish meiosis II, the four resulting cells are haploid spermatids that will each mature into a sperm cell, a spermatozoon.

During oogenesis, diploid cells in the ovaries called oogonia divide by mitosis. Some of the resulting cells differentiate into primary oocytes, which divide by meiosis. Meiosis I produces two haploid cells, which receive unequal amounts of cytoplasm. The cell that receives most of the cytoplasm is the secondary oocyte; the small cell is called a polar body. The polar body completes meiosis to produce two polar bodies. When the secondary oocyte undergoes meiosis II, the cytoplasm is again divided unequally, producing a haploid ootid with most of the cytoplasm and another smaller polar body. The ootid matures into an ovum, and the three polar bodies degenerate.

Gametogenesis in Animals

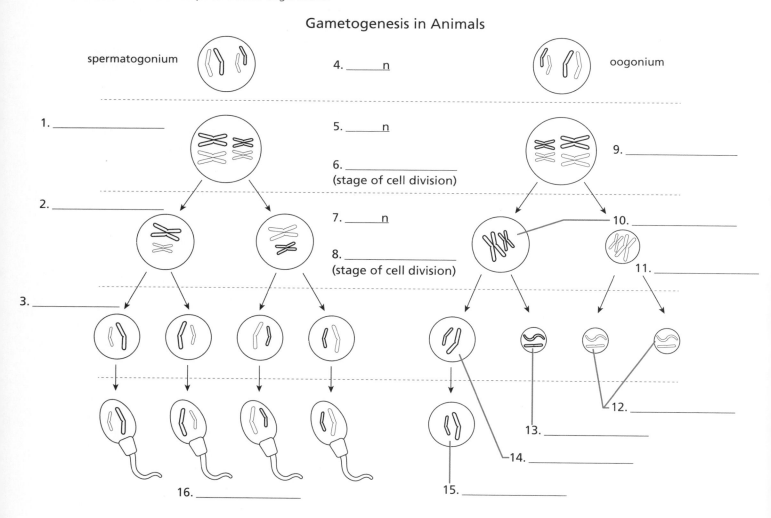

Answers

Respiratory System

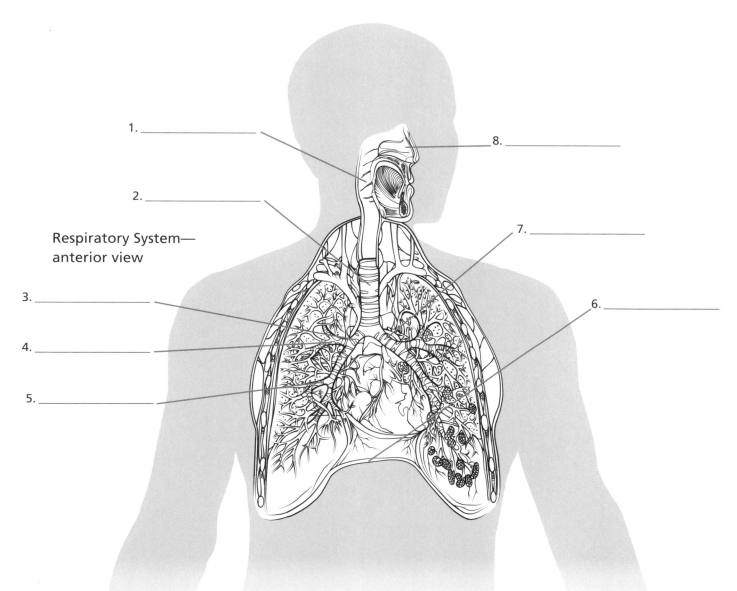

1. _____

2. _____

Respiratory System—
anterior view

3. _____

4. _____

5. _____

8. _____

7. _____

6. _____

The respiratory system brings oxygen into the body and then transfers it to the blood, where it can be carried around the body to all cells. Cells use oxygen during cellular respiration, which transfers energy from food to ATP. In the process, cells produce carbon dioxide as waste. This gas is transferred to the blood and then to the respiratory system to be expelled from the body.

The respiratory system consists of your nose, throat (pharynx), trachea, lungs, and diaphragm. As your trachea descends into your lungs, it divides into two large branches called bronchi, which split into smaller and smaller branches until they become minute passageways called bronchioles. At the tips of the bronchioles are small sacs called alveolar sacs. Each of these sacs contains many chambers called alveoli, which are surrounded by blood capillaries.

The movement of air in and out of your lungs is called ventilation. Your diaphragm contracts and moves downward, creating negative pressure inside your lungs. Air moves into your body and through your air passages until it reaches the alveoli. Gases are exchanged between the alveoli and the capillaries. Your diaphragm then relaxes and moves up, increasing pressure in your lungs so that air exits your body.

Answers

1. pharynx, 2. trachea, 3. right primary bronchus, 4. superior lobar bronchus, 5. middle lobar bronchus, 6. diaphragm, 7. left primary bronchus, 8. nasal cavity

Digestive System

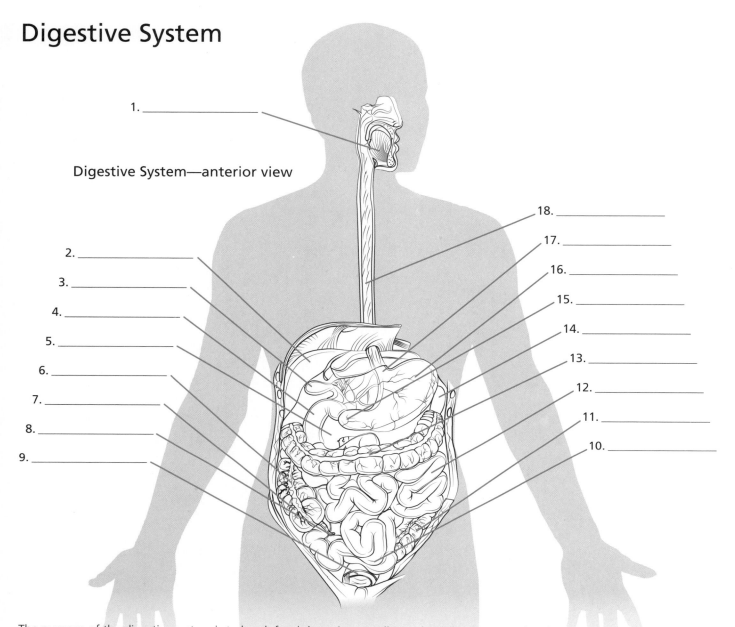

Digestive System—anterior view

1. _____

2. _____
3. _____
4. _____
5. _____
6. _____
7. _____
8. _____
9. _____

18. _____
17. _____
16. _____
15. _____
14. _____
13. _____
12. _____
11. _____
10. _____

The purpose of the digestive system is to break food down into smaller and smaller pieces so that food molecules can be absorbed by cells in the small intestine. After food is absorbed, waste materials pass into the large intestine, which reabsorbs water and minerals before expelling waste through the anus.

The digestive system consists of a long tube called the alimentary tract, plus the liver, gallbladder, and pancreas, which are called accessory organs. The alimentary tract runs from the mouth to the anus and includes the pharynx, esophagus, stomach, and small and large intestines. The small intestine has three regions: the upper duodenum, the middle jejunum, and the lower ileum. The accessory organs have ducts that open into the alimentary tract.

The digestion of food occurs by both mechanical and chemical processes. Mechanical digestion begins in the mouth with the physical activity of teeth and tongue and continues in the stomach with the churning of stomach contents. Chemical digestion occurs through enzyme action, beginning in the mouth and continuing in the stomach and small intestine. Digestion in the small intestine is aided by secretions from the pancreas and gallbladder that enter the duodenum. Bile, which is produced by the liver and secreted by the gallbladder, emulsifies fats. Pancreatic juice contains a mixture of digestive enzymes.

Answers

1. tongue, 2. liver (lifted up), 3. gallbladder, 4. duodenum, 5. pancreas, 6. ascending colon, 7. cecum, 8. appendix, 9. rectum, 10. ileum, 11. descending colon, 12. jejunum, 13. transverse colon, 14. spleen, 15. pylorus, 16. stomach, 17. gastroesophageal (cardioesophageal) junction, 18. esophagus

Urinary System

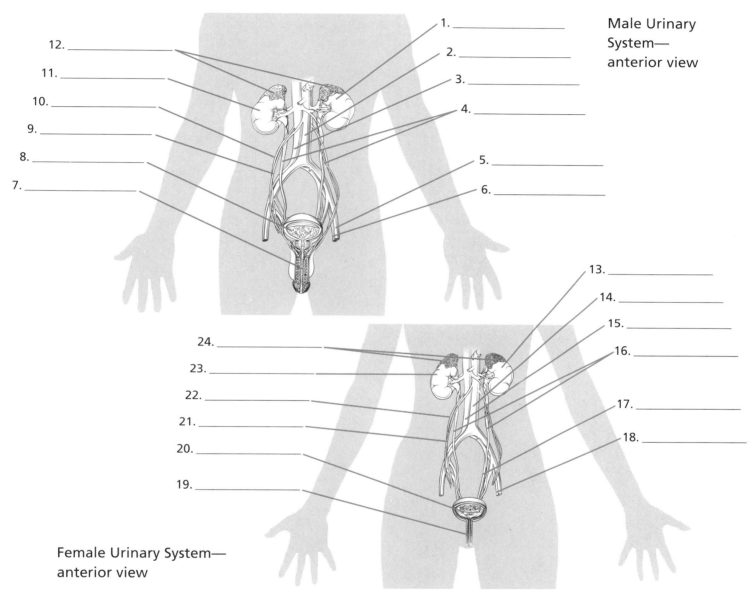

12. _____
11. _____
10. _____
9. _____
8. _____
7. _____

1. _____
2. _____
3. _____
4. _____
5. _____
6. _____

Male Urinary System— anterior view

24. _____
23. _____
22. _____
21. _____
20. _____
19. _____

13. _____
14. _____
15. _____
16. _____
17. _____
18. _____

Female Urinary System— anterior view

The primary purpose of the urinary system is to eliminate metabolic wastes from the body. As cells perform metabolic processes, they release wastes such as urea into the blood. The blood passes through the kidneys, which filter out the wastes and combine them with salt and water to form urine. The kidneys also maintain homeostasis in the blood by regulating water and electrolyte concentration as well as blood pH. Urine travels from the kidneys to the urinary bladder before being excreted from the body.

The urinary system consists of the bladder, ureters, kidneys, and urethra. Urine passes from the kidneys to the bladder via the ureters. The bladder acts as the holding tank for urine. As the bladder fills, pressure from urine activates pressure receptors that send impulses to your brain. Once the bladder is full, the brain sends an impulse to relax the sphincter that holds urine in the bladder, causing the excretion of urine via the urethra. In males, the urethra passes through the pelvic floor and into the penis, where it serves as the passageway for both urine and sperm. In females, the urethra is very short and opens in front of the entry to the vagina.

Answers

1. left kidney, 2. abdominal aorta, 3. inferior vena cava, 4. ureters, 5. external iliac artery, 6. external iliac vein, 7. urethra, 8. bladder, 9. testicular artery, 10. testicular vein, 11. right kidney, 12. adrenal glands, 13. left kidney, 14. abdominal aorta, 15. inferior vena cava, 16. ureters, 17. external iliac artery, 18. external iliac vein, 19. urethra, 20. bladder, 21. ovarian artery, 22. ovarian vein, 23. right kidney, 24. adrenal glands

Close-Up on a Nephron

The kidneys of healthy adults contain more than a million nephrons. Each nephron consists of a renal corpuscle and a renal tubule. The renal corpuscle contains a cup-shaped structure called the Bowman's capsule, which surrounds a tuft of capillaries called the glomerulus. The renal tubule has three sections: the proximal convoluted tubule, which connects to the Bowman's capsule; the loop of Henle, which descends and ascends through the medulla of the kidney; and the distal convoluted tubule, which connects to the collecting duct.

Each part of the nephron contributes to the filtration of blood and reabsorption of useful materials. Filtration occurs as pressure within the capillaries of the glomerulus forces water and salts through the Bowman's capsule and into the proximal tubule. Cells that line the proximal tubule selectively reabsorb salt, water, and food molecules, transferring them to nearby capillaries. The fluid flows into the descending portion of the loop of Henle, losing water by osmosis to the hypertonic renal medulla. The fluid travels through the ascending loop of Henle, where cells actively pump out ions such as sodium. The cells that form the ascending loop are impermeable to water. As the fluid passes through the distal convoluted tubule, ions are actively transported in and out of the fluid in response to body signals. Urine passes into the collecting duct, which is normally impermeable to water but can adjust its permeability in order to maintain blood pressure.

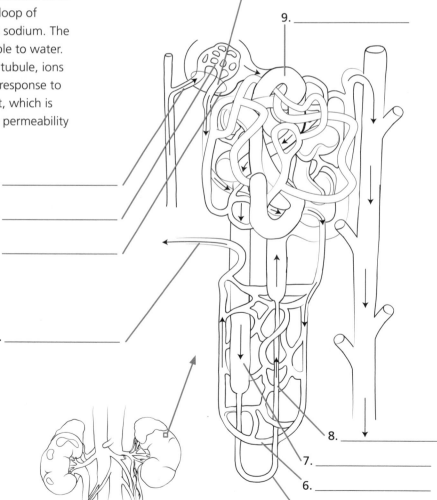

10. _____

9. _____

1. _____

2. _____

3. _____

Nephron

4. _____

8. _____

7. _____

6. _____

5. _____

Circulatory System

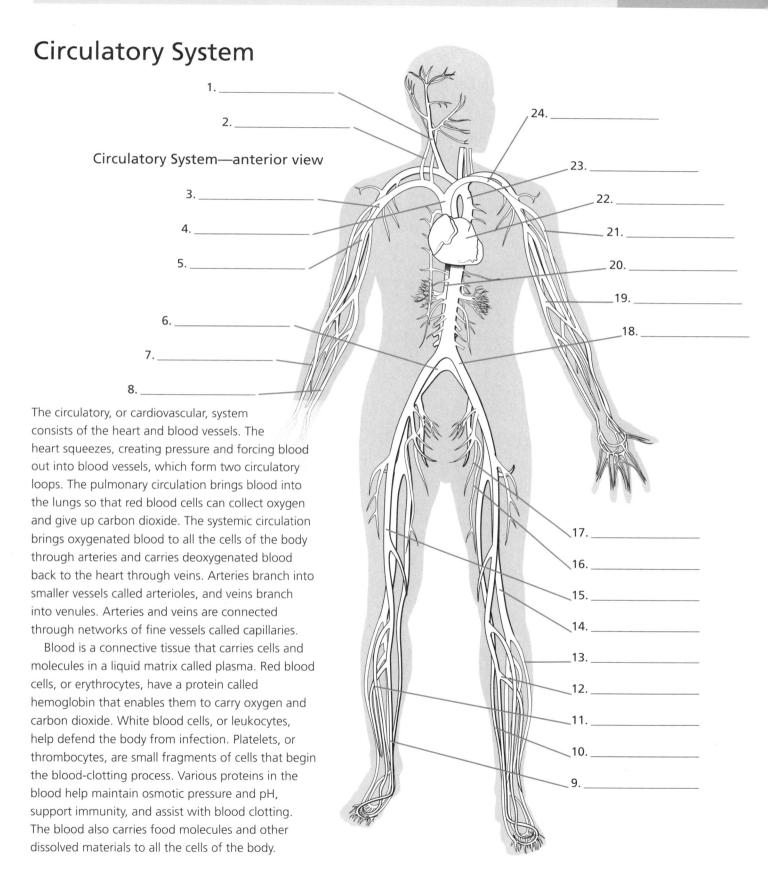

Circulatory System—anterior view

1. _____

2. _____

3. _____

4. _____

5. _____

6. _____

7. _____

8. _____

24. _____

23. _____

22. _____

21. _____

20. _____

19. _____

18. _____

17. _____

16. _____

15. _____

14. _____

13. _____

12. _____

11. _____

10. _____

9. _____

The circulatory, or cardiovascular, system consists of the heart and blood vessels. The heart squeezes, creating pressure and forcing blood out into blood vessels, which form two circulatory loops. The pulmonary circulation brings blood into the lungs so that red blood cells can collect oxygen and give up carbon dioxide. The systemic circulation brings oxygenated blood to all the cells of the body through arteries and carries deoxygenated blood back to the heart through veins. Arteries branch into smaller vessels called arterioles, and veins branch into venules. Arteries and veins are connected through networks of fine vessels called capillaries.

Blood is a connective tissue that carries cells and molecules in a liquid matrix called plasma. Red blood cells, or erythrocytes, have a protein called hemoglobin that enables them to carry oxygen and carbon dioxide. White blood cells, or leukocytes, help defend the body from infection. Platelets, or thrombocytes, are small fragments of cells that begin the blood-clotting process. Various proteins in the blood help maintain osmotic pressure and pH, support immunity, and assist with blood clotting. The blood also carries food molecules and other dissolved materials to all the cells of the body.

Answers

1. common carotid artery, 2. external jugular vein, 3. axillary vein, 4. superior vena cava, 5. brachial artery, 6. common iliac vein, 7. radial artery, 8. ulnar artery, 9. posterior tibial artery, 10. great saphenous vein, 11. anterior tibial artery, 12. fibular artery, 13. small saphenous vein, 14. popliteal vein, 15. femoral artery, 16. obturator artery, 17. obturator vein, 18. common iliac artery, 19. basilic vein, 20. inferior vena cava, 21. cephalic vein, 22. heart, 23. arch of aorta, 24. subclavian vein

Anatomy of the Human Heart

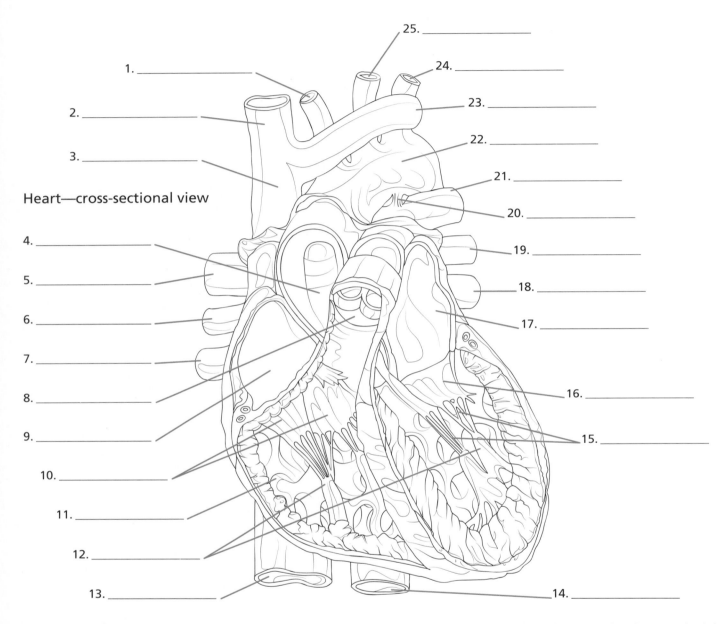

Heart—cross-sectional view

25. _____

24. _____

23. _____

22. _____

21. _____

20. _____

19. _____

18. _____

17. _____

16. _____

15. _____

1. _____

2. _____

3. _____

4. _____

5. _____

6. _____

7. _____

8. _____

9. _____

10. _____

11. _____

12. _____

13. _____

14. _____

The heart is divided down the middle by a septum and into upper and lower chambers by valves. The upper chambers are the left and right atria; the lower chambers are the left and right ventricles. Oxygenated blood from the lungs enters the left atrium through the pulmonary veins, while deoxygenated blood from the body enters the right atrium through the superior and inferior vena cava. Blood drains from the atria into their respective ventricles, helped by atrial contraction. The ventricles contract to push blood from the left ventricle out through the aorta and into the systemic circulation, and from the right ventricle out through the pulmonary artery and into the pulmonary circulation. When the ventricles contract, valves close to prevent the backwards flow of blood. First, the valves between the atria and ventricles snap shut to make the "lub" sound of your heartbeat; these are the mitral valve on the left and the tricuspid valve on the right. As blood exits the heart, the aortic and pulmonary valves snap shut to prevent backflow from the aorta and pulmonary artery, respectively, making the "dub" sound.

Answers

1. brachiocephalic trunk, 2. right brachiocephalic vein, 3. superior vena cava, 4. ascending aorta, 5. right pulmonary artery, 6. right superior pulmonary vein, 7. right inferior pulmonary vein, 8. pulmonary valve, 9. right atrium, 10. cusp of tricuspid valve, 11. right ventricle, 12. papillary muscles, 13. inferior vena cava, 14. descending thoracic aorta, 15. chordae tendineae, 16. cusp of mitral valve, 17. left atrium, 18. left inferior pulmonary vein, 19. left superior pulmonary vein, 20. ligamentum arteriosum, 21. left pulmonary artery, 22. aortic arch, 23. left brachiocephalic vein, 24. left subclavian artery, 25. left common carotid artery

Major Skeletal Muscles

Muscular System—
anterior view

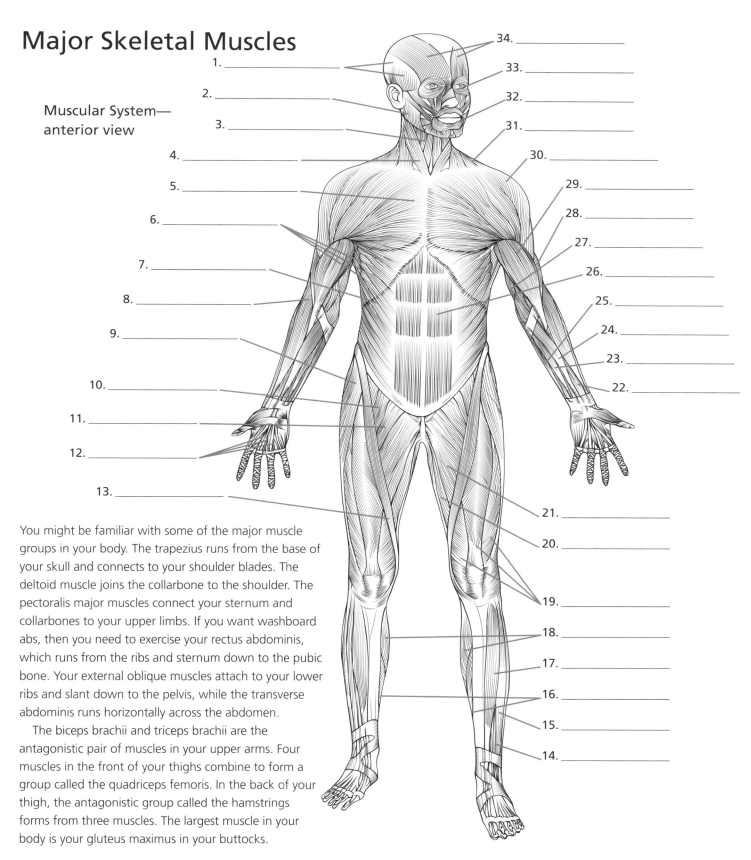

1. _____
2. _____
3. _____
4. _____
5. _____
6. _____
7. _____
8. _____
9. _____
10. _____
11. _____
12. _____
13. _____

34. _____
33. _____
32. _____
31. _____
30. _____
29. _____
28. _____
27. _____
26. _____
25. _____
24. _____
23. _____
22. _____
21. _____
20. _____
19. _____
18. _____
17. _____
16. _____
15. _____
14. _____

You might be familiar with some of the major muscle groups in your body. The trapezius runs from the base of your skull and connects to your shoulder blades. The deltoid muscle joins the collarbone to the shoulder. The pectoralis major muscles connect your sternum and collarbones to your upper limbs. If you want washboard abs, then you need to exercise your rectus abdominis, which runs from the ribs and sternum down to the pubic bone. Your external oblique muscles attach to your lower ribs and slant down to the pelvis, while the transverse abdominis runs horizontally across the abdomen.

The biceps brachii and triceps brachii are the antagonistic pair of muscles in your upper arms. Four muscles in the front of your thighs combine to form a group called the quadriceps femoris. In the back of your thigh, the antagonistic group called the hamstrings forms from three muscles. The largest muscle in your body is your gluteus maximus in your buttocks.

Answers

1. temporalis, 2. masseter, 3. sternohyoid, 4. sternocleidomastoid, 5. pectoralis major, 6. serratus anterior, 7. external oblique, 8. brachioradialis, 9. brachioradialis, 10. iliopsoas, 11. pectineus, 12. lumbricals, 13. sartorius, 14. extensor hallucis longus, 15. extensor digitorum longus, 16. soleus, 17. tibialis anterior, 18. gastrocnemius, 19. quadriceps femoris, 20. adductor magnus, 21. adductor longus, 22. flexor digitorum superficialis, 23. palmaris longus, 24. flexor carpi radialis, 25. flexor carpi ulnaris, 26. rectus abdominis, 27. triceps brachii, 28. brachialis, 29. biceps brachii, 30. deltoid, 31. trapezius, 32. orbicularis oris, 33. orbicularis oculi, 34. frontalis

Structure and Function of Skeletal Muscle

Muscle tissue consists of bundles of long tubular muscle cells called muscle fibers. Each muscle fiber contains many nuclei and is stuffed with bands of linear proteins called myofibrils. Each myofibril is a bundle of cytoskeletal proteins arranged into alternating layers of thin filaments (actin) and thick filaments (myosin), together called actomyosin. Thick filaments have a bulbous head that can attach and release from binding sites on the thin filaments. Proteins that attach neighboring groups of thin filaments form dark lines in the muscle called Z lines. The area of actomyosin between two Z lines is called a sarcomere; each sarcomere acts as an independent contractile unit in the muscle.

When you send a signal from your nervous system, your motor neurons trigger the release of calcium from the sarcoplasmic reticulum of the muscle cells. Calcium floods the muscle cells, moving a regulatory protein (tropomyosin) and exposing binding sites on the thin filaments. Using energy from ATP, the thick filaments bind to the thin filaments and pull on them, contracting the sarcomere. The thick filaments release their grip and then repeat the action using more ATP. As long as there's enough ATP and a signal from your motor neurons, this process will continue until the muscle is fully contracted.

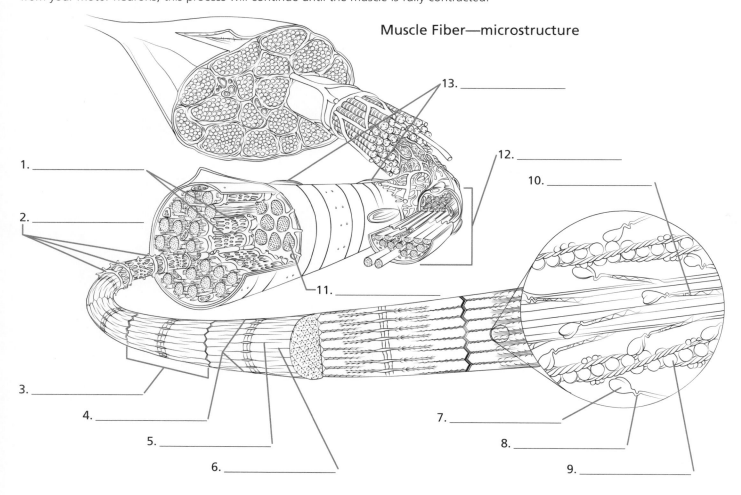

Muscle Fiber—microstructure

13. _____

12. _____

10. _____

1. _____

2. _____

11. _____

3. _____

4. _____

5. _____

6. _____

7. _____

8. _____

9. _____

Major Bones

Skeletal System—anterior view

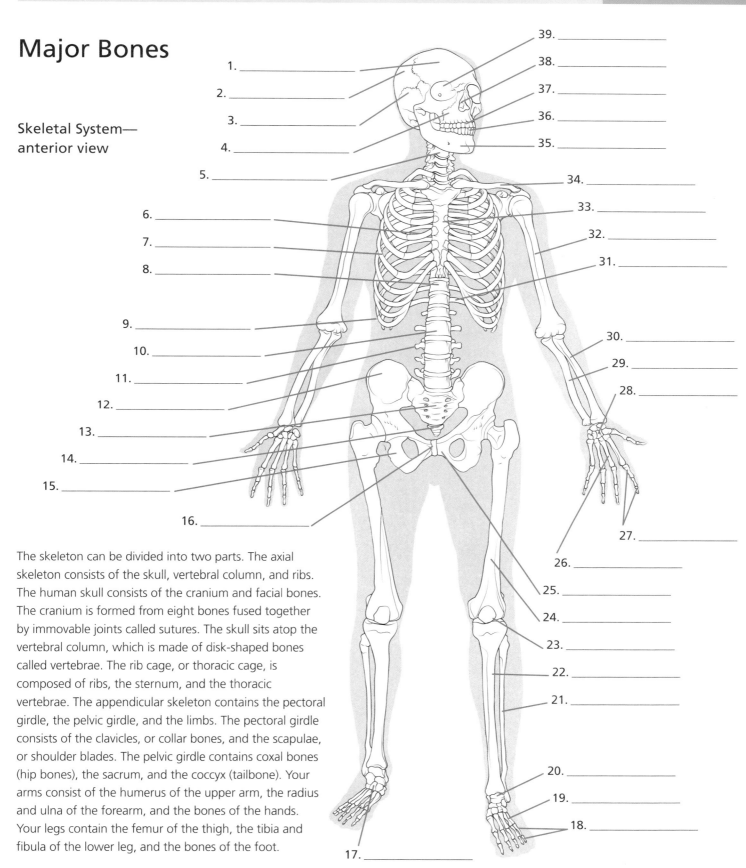

1. _____
2. _____
3. _____
4. _____
5. _____
6. _____
7. _____
8. _____
9. _____
10. _____
11. _____
12. _____
13. _____
14. _____
15. _____
16. _____
17. _____
18. _____
19. _____
20. _____
21. _____
22. _____
23. _____
24. _____
25. _____
26. _____
27. _____
28. _____
29. _____
30. _____
31. _____
32. _____
33. _____
34. _____
35. _____
36. _____
37. _____
38. _____
39. _____

The skeleton can be divided into two parts. The axial skeleton consists of the skull, vertebral column, and ribs. The human skull consists of the cranium and facial bones. The cranium is formed from eight bones fused together by immovable joints called sutures. The skull sits atop the vertebral column, which is made of disk-shaped bones called vertebrae. The rib cage, or thoracic cage, is composed of ribs, the sternum, and the thoracic vertebrae. The appendicular skeleton contains the pectoral girdle, the pelvic girdle, and the limbs. The pectoral girdle consists of the clavicles, or collar bones, and the scapulae, or shoulder blades. The pelvic girdle contains coxal bones (hip bones), the sacrum, and the coccyx (tailbone). Your arms consist of the humerus of the upper arm, the radius and ulna of the forearm, and the bones of the hands. Your legs contain the femur of the thigh, the tibia and fibula of the lower leg, and the bones of the foot.

Answers

1. frontal bone, 2. parietal bone, 3. temporal bone, 4. maxilla, 5. cervical vertebra, 6. costal cartilage, 7. true rib, 8. thoracic vertebra, 9. false rib, 10. lumbar vertebra, 11. transverse process, 12. ilium, 13. sacrum, 14. coccyx, 15. ischium, 16. symphysis pubis, 17. tarsal bones, 18. phalanges, 19. metatarsal bones, 20. talus, 21. fibula, 22. tibia, 23. patella, 24. femur, 25. pubic bone, 26. metacarpal bones, 27. phalanges, 28. carpal bones, 29. ulna, 30. radius, 31. twelfth rib (floating rib), 32. humerus, 33. sternum, 34. clavicle, 35. mandible, 36. lower teeth, 37. upper teeth, 38. anterior nasal aperture, 39. orbit

Nervous System

The nervous system receives and processes information, controls movement, stores learning, and regulates the organs. The central nervous system consists of the brain and the spinal cord. The peripheral nervous system carries information to and from the central nervous system. The functional unit of the nervous system is the neuron, or nerve cell, which generates and conducts signals called nerve impulses. Each impulse is based on the movement of ions across the plasma membrane of the neuron. Afferent, or sensory, neurons carry information from sensory receptors to the central nervous system. Efferent neurons carry information from the central nervous system to effectors such as muscles and glands.

The peripheral nervous system can be divided into the somatic (or voluntary) system, which is responsible for control of body movements, and the autonomic system, which automatically controls organs, glands, and smooth muscle. The autonomic nervous system can be further divided into the sympathetic and parasympathetic systems. The sympathetic nervous system is sometimes called the "fight or flight" system because it can lead to the release of epinephrine (adrenaline), resulting in increased heart rate, pupil dilation, sweating, and blood pressure. The parasympathetic system is referred to as the "rest and digest" system because it slows the heart rate and stimulates digestion.

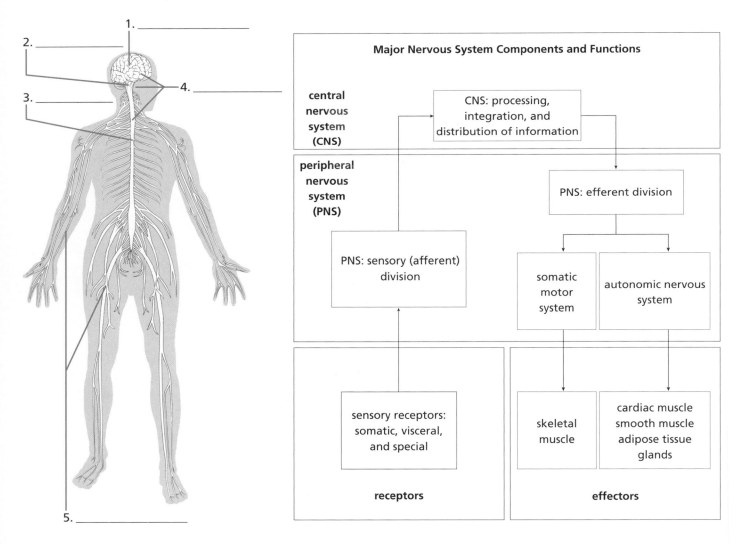

1. _____
2. _____
3. _____
4. _____
5. _____

Major Nervous System Components and Functions

central nervous system (CNS)

CNS: processing, integration, and distribution of information

peripheral nervous system (PNS)

PNS: efferent division

PNS: sensory (afferent) division

somatic motor system

autonomic nervous system

sensory receptors: somatic, visceral, and special

skeletal muscle

cardiac muscle smooth muscle adipose tissue glands

receptors

effectors

Answers

Close-Up of Neurons and Action Potentials

Neurons consist of a cell body, which contains the nucleus and organelles, and extensions called dendrites and axons. Dendrites receive information and transmit impulses toward the cell body, while axons send impulses away. Axons may be covered with a fatty layer of insulation called the myelin sheath. A neuron that's transmitting a signal releases chemicals called neurotransmitters from the synaptic terminals of its axon. The neurotransmitters move into the space between the neurons, called the synaptic cleft, and bind to receptors on dendrites of the receiving cell.

The electrical changes that occur in a neuron in response to a signal are called an action potential. Before a neuron receives a signal, the inside of the cell is more negative than the outside. This charge differential, or polarization, is called the resting potential of the cell. It's set up by a protein called the sodium-potassium pump, which constantly moves three sodium ions (Na^+) outside the cell for every two potassium ions (K^+) it brings in.

When a neuron receives a signal, voltage-gated sodium channels open in the membrane and allow sodium to diffuse into the cell, causing the membrane to depolarize. After the inside of the cell is flooded with sodium ions, voltage-gated potassium channels open and the sodium channels close. Potassium diffuses out of the cell, repolarizing the neuron and eventually leading to hyperpolarization. The potassium channels close, and the sodium-potassium pump returns the sodium and potassium ions to their original distribution.

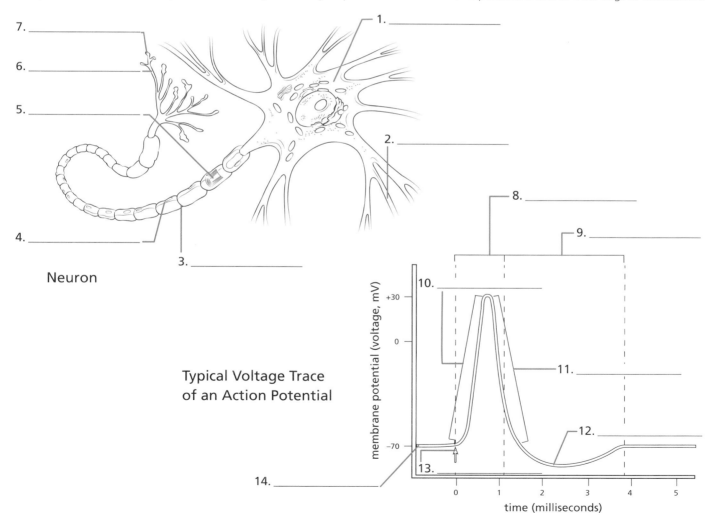

Neuron

Typical Voltage Trace
of an Action Potential

Answers

The Human Eye

The eye focuses and collects visual information and then transmits it to the brain along the optic nerve. The outer layer of the eye consists of the cornea and sclera. The cornea is the clear area that covers the front of the eye; the sclera is the white protective layer that gives the eyeball its shape.

The lens and uvea—which consists of the choroid, ciliary body, and iris—are the middle layer of the eye. The iris is the colored muscular part of the eye that contracts and expands in order to adjust the size of the pupil and let in the correct amount of light. The transparent lens focuses light onto the retina at the back of the eye. The choroid is the vascular layer of the eye. The ciliary body contains the ciliary muscle, which controls the shape of the lens, and the ciliary epithelium, which produces aqueous humor. Aqueous humor is a transparent watery fluid that fills the anterior cavity in front of the lens.

The posterior cavity of the eye is the space between the lens and the retina. This cavity is filled with vitreous humor, a transparent gel. The retina contains two types of photoreceptors called rods and cones. Rods enable vision in dim light. Cones require higher amounts of light but can detect color.

Eye—lateral view

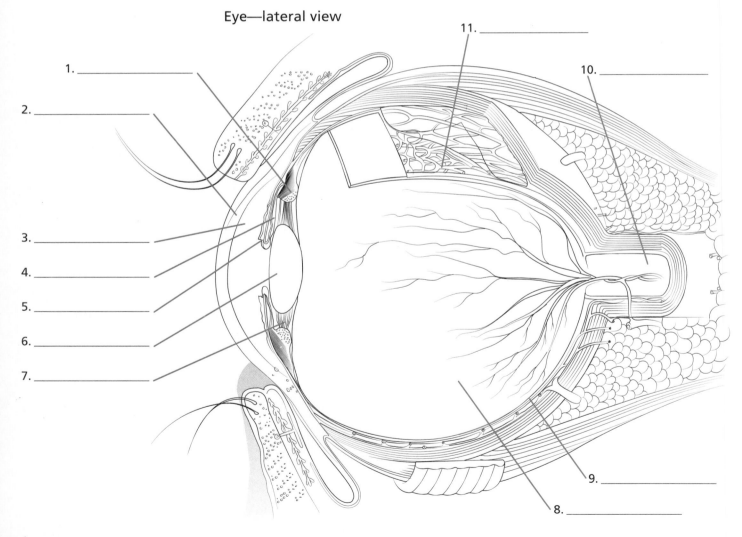

11. _____

1. _____

2. _____

10. _____

3. _____

4. _____

5. _____

6. _____

7. _____

9. _____

8. _____

Answers

1. ciliary body, 2. cornea, 3. anterior chamber (of anterior cavity), 4. posterior chamber (of posterior cavity), 5. iris, 6. lens, 7. ciliary fibers, 8. vitreous body (of posterior cavity), 9. retina, 10. optic nerve, 11. choroid.

The Human Ear

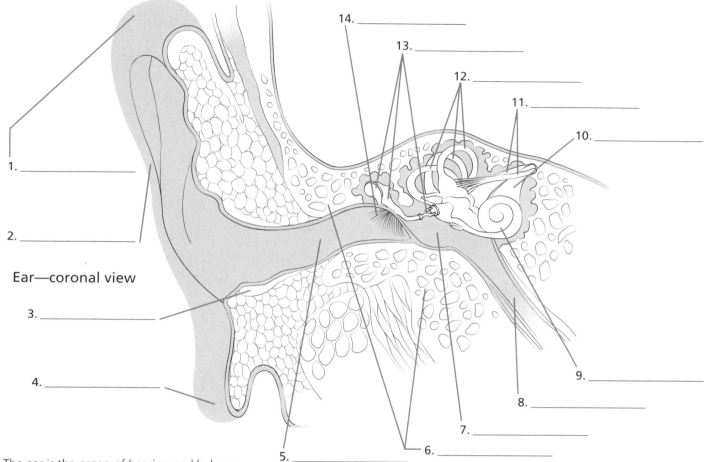

Ear—coronal view

1. _____

2. _____

3. _____

4. _____

5. _____

6. _____

7. _____

8. _____

9. _____

10. _____

11. _____

12. _____

13. _____

14. _____

The ear is the organ of hearing and balance. The external part of the ear, called the auricle or pinna, acts as a funnel to bring sound to the tympanic membrane (eardrum). Sound energy strikes the membrane and causes it to vibrate. These vibrations pass into the middle ear, causing small bones called ossicles to vibrate. The three ossicles in order from outside to inside are the malleus, incus, and stapes. The ossicles transmit the vibrations to the oval window at the boundary of the inner ear.

The inner ear consists of a series of chambers and passages in the temporal bone of the skull, which scientists call the bony labyrinth. The bony labyrinth has two main parts: the cochlea, which processes sound information and sends it to the brain, and the vestibular system, which maintains balance.

Tympanic Membrane (Eardrum)—internal view

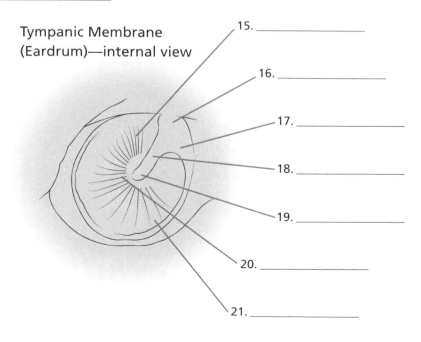

15. _____

16. _____

17. _____

18. _____

19. _____

20. _____

21. _____

Answers

The Human Nose

The nose is the organ of smell, or olfaction. The upper chamber of the nose is lined with olfactory cells, which have chemoreceptors to detect molecules in the air you breathe. Air enters through your nasal vestibule, or nostrils. The chemicals diffuse into the mucus that lines your nasal cavity and bind to the cilia of the olfactory cells. The olfactory cells trigger signals in sensory neurons that pass through the ethmoid bone to the olfactory bulbs. The olfactory bulbs are oval-shaped extensions of the olfactory center of the brain that contain several different types of neurons. The axons of the sensory neurons synapse with the dendrites of the neurons in the olfactory bulbs. The axons of these neurons form olfactory tracts, which travel to the olfactory areas of the brain that interpret olfactory information.

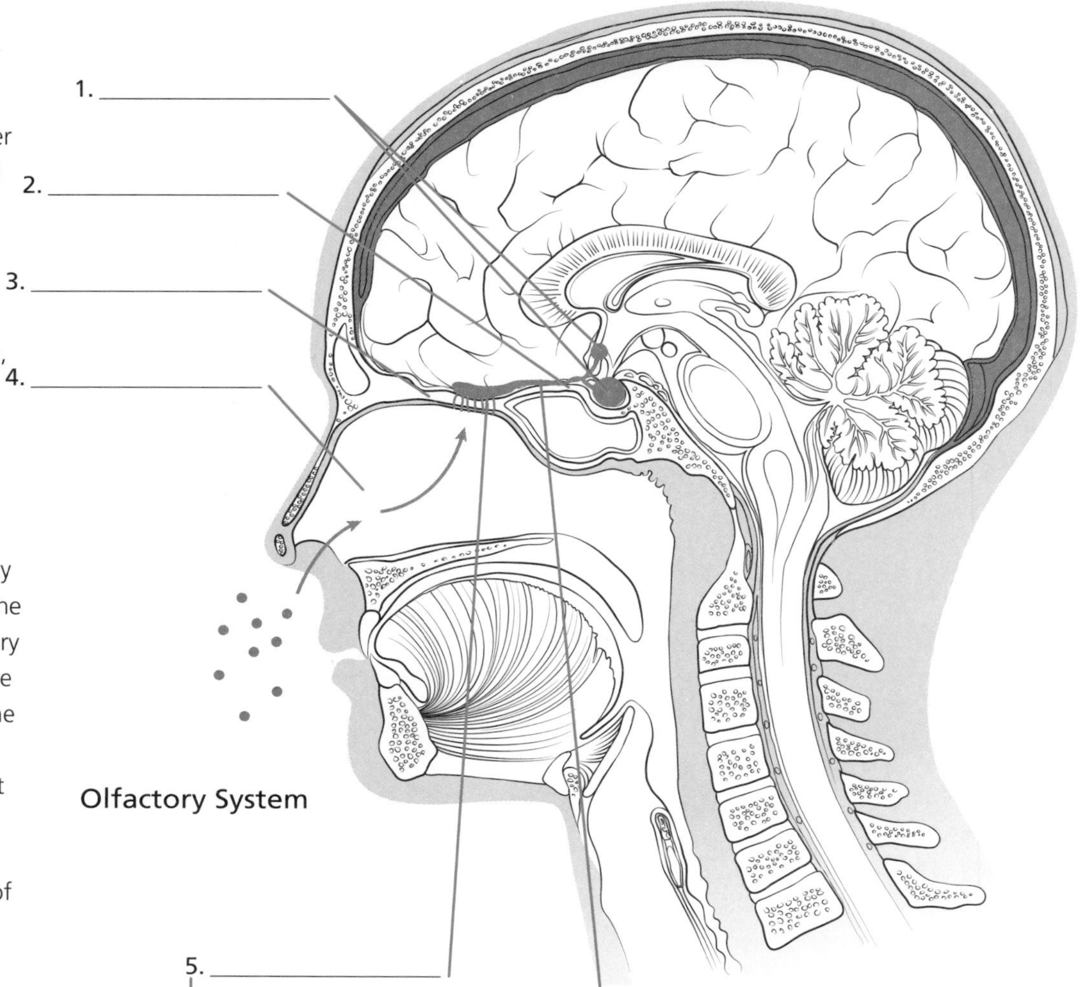

1. _____
2. _____
3. _____
4. _____

Olfactory System

5. _____
6. _____
7. _____
8. _____
9. _____
10. _____
11. _____
12. _____

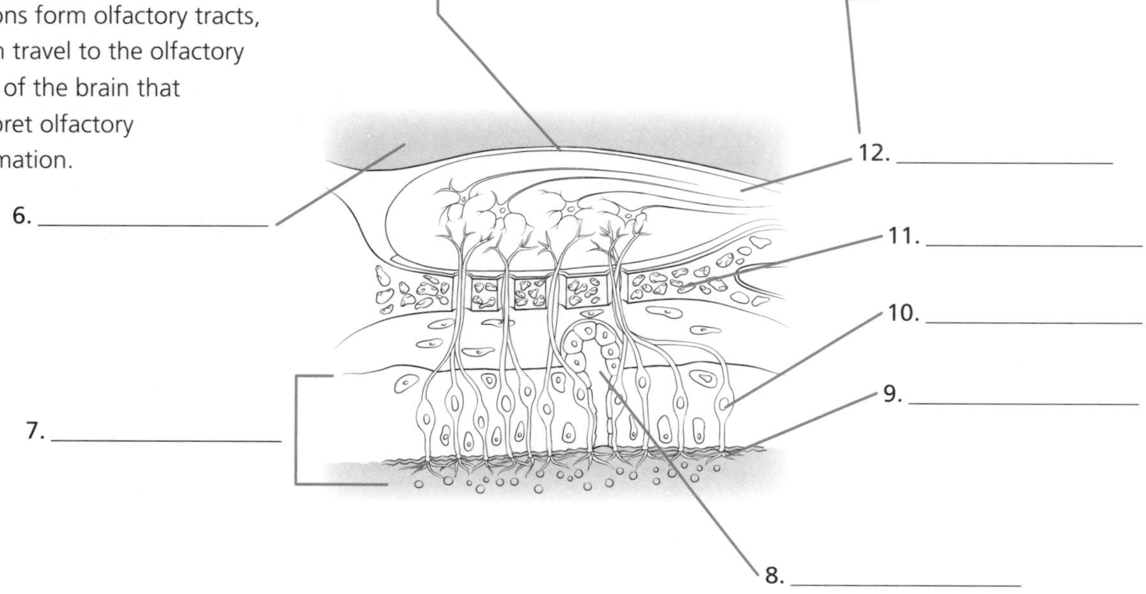

Answers

1. olfactory centers in the brain (in septal region and temporal lobe), 2. sphenoid bone, 3. olfactory receptors, 4. nasal cavity, 5. olfactory bulb, 6. frontal lobe of the brain, 7. olfactory mucosa, 8. Bowman's gland (olfactory gland), 9. cilia, 10. olfactory cell, 11. cribriform plate of the ethmoid bone, 12. olfactory tract

The Tongue

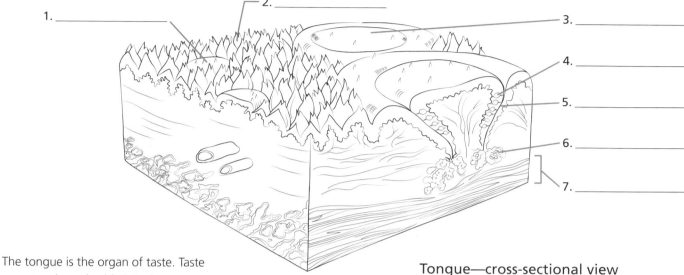

1. _____
2. _____
3. _____
4. _____
5. _____
6. _____
7. _____

Tongue—cross-sectional view

The tongue is the organ of taste. Taste receptors (taste buds) sit in the grooves around the bumps, or papillae, on your tongue, as well as on the roof of the mouth, the epiglottis, and the entrance to the pharynx. Your tongue has three different types of papillae: club-shaped fungiform papillae, cone-shaped filiform papillae, and dome-shaped circumvallate (vallate) papillae. Each taste bud consists of several sensory receptor cells surrounded by supporting cells. Dissolved food material passes through a pore canal in the top of the taste bud and binds to receptors within the bud, triggering nerve impulses that transmit through the cranial nerves to the medulla in the brainstem, on to the thalamus, and ultimately to the taste areas in the parietal lobes of the cerebral cortex.

Taste buds can detect five different types of taste: sweet, salty, bitter, sour, and umami (savory). Each type of flavor activates a different set of taste receptors. For many years, people have misinterpreted a map of the tongue to mean that these different receptors are localized in different parts of the tongue. Scientists have shown, however, that receptors for all the tastes occur in all parts of the tongue.

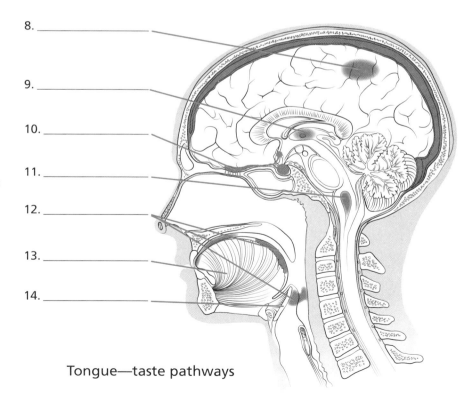

8. _____
9. _____
10. _____
11. _____
12. _____
13. _____
14. _____

Tongue—taste pathways

Answers

Endocrine System

The endocrine system produces hormones, chemical messengers that are produced by endocrine cells and then travel some distance through the body before they regulate the action of other cells and tissues. Many hormones, such as estrogen and testosterone, are steroid hormones. Nonsteroid hormones, such as insulin and epinephrine, are proteins or similar molecules. When steroid hormones reach their target cell, they pass through the plasma membrane and bind to receptors in the cell cytoplasm. Nonsteroid hormones bind to receptors in the plasma membrane and transmit their signals via signal transduction. The endocrine system works with the nervous system to control the body and maintain homeostasis.

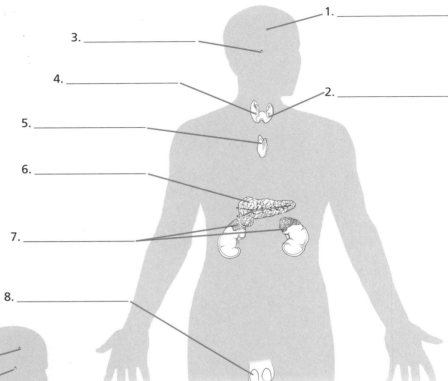

1. _____
2. _____
3. _____
4. _____
5. _____
6. _____
7. _____
8. _____

Endocrine System (male)— anterior view

9. _____
10. _____
11. _____
12. _____
13. _____
14. _____
15. _____
16. _____

Endocrine System (female)— anterior view

Men and women have many of the same endocrine organs, including the pituitary gland, pineal gland, thymus, thyroid, parathyroid glands, adrenal glands, and pancreatic islets. The endocrine organs unique to men are the testes, and those unique to women are the ovaries and placenta (during pregnancy). The testes and ovaries produce hormones that regulate sexual function and trigger the development of secondary sex characteristics, such as facial hair and an Adam's apple in men and enlarged breasts and widened hips in women.

Answers

1. pineal gland, 2. thyroid gland, 3. pituitary gland, 4. parathyroid, 5. thymus, 6. pancreas, 7. adrenal glands, 8. testes, 9. pineal gland, 10. pituitary gland, 11. parathyroid, 12. thyroid gland, 13. thymus, 14. pancreas, 15. adrenal glands, 16. ovaries

Male Reproductive System

The biological function of the male reproductive system is to produce and transmit sperm to a female so that fertilization of an egg can occur. The external structures of the male reproductive system are the penis and the scrotum. The scrotum contains two testes, the gonads that produce sperm and male hormones. During ejaculation, sperm leaves each testis through a tube called the epididymis and travels to the prostate gland via the vas deferens. It combines with secretions from the prostate gland and seminal vesicle to form seminal fluid (semen). This fluid then exits the penis via the urethra.

The production of sperm occurs inside the testes, which are divided into compartments by fibrous tissue. The compartments are filled with coiled seminiferous tubules. Spermatogenesis begins with the spermatogonia that line the walls of the seminiferous tubules. Other cells in the tubules, called Sertoli cells, provide food for the developing sperm. Mature sperm can survive inside the epididymis and vas deferens for up to six weeks.

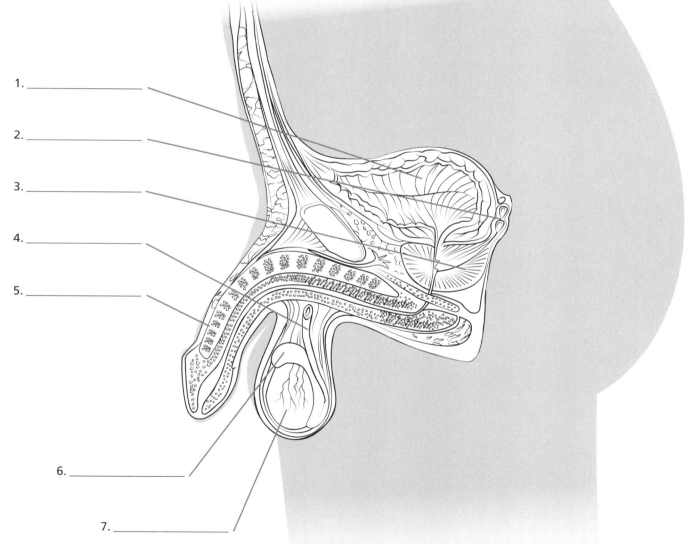

1. _____

2. _____

3. _____

4. _____

5. _____

6. _____

7. _____

Male Reproductive System—
sagittal view

Answers

1. urinary bladder, 2. seminal vesicle, 3. prostate gland, 4. vas deferens/ductus deferens, 5. penis, 6. epididymis, 7. testis

Female Reproductive System

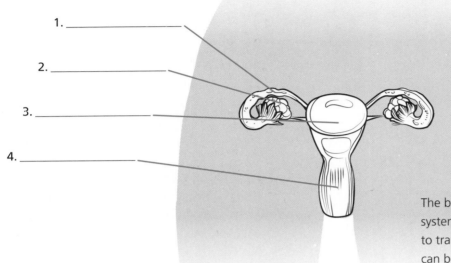

1. _____

2. _____

3. _____

4. _____

Female Reproductive System —anterior view

Female Reproductive System— sagittal view

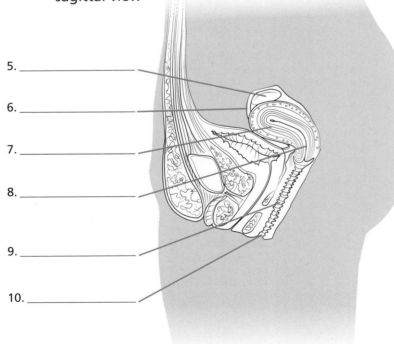

5. _____

6. _____

7. _____

8. _____

9. _____

10. _____

The biological function of the female reproductive system is to produce ova (eggs) for fertilization and to transport fertilized ova to the uterus, where they can be supported until birth. The external part of the female system is the vulva, which consists of the labia majora, labia minora, and clitoris. The labia are the folds of tissue that protect the opening of the vagina; the labia majora are the external folds, while the labia minora are the internal folds. The clitoris is a bud of tissue that is sensitive to sexual stimulation.

The internal structures of the female system are the ovaries, fallopian tubes, uterus, and vagina. The ovaries are the gonads that produce ova and female hormones. The ova travel through the fallopian tubes to the muscular uterus. The uterine lining, called the endometrium, thickens and sheds during the menstrual cycle. The lower third of the uterus, called the cervix, opens into the vagina, a muscular tube that extends to the vulva.

The production of ova occurs inside the ovaries. Early in the development of a female fetus, oogonia in the ovaries undergo mitosis to produce primary oocytes. Groups of cells surround these oocytes to form multilayered follicles within the ovary. The primary oocytes begin meiosis but arrest in prophase I and remain arrested until the female reaches sexual maturity. At that point, female hormones, produced during the menstrual cycle, stimulate the development of follicles and the completion of meiosis to produce ova.

Answers

1. fallopian tube, 2. ovary, 3. uterus, 4. vagina, 5. ovary, 6. fallopian tube, 7. uterus, 8. cervix, 9. vagina, 10. vaginal opening to vulva

Menstrual Cycle

The menstrual cycle is the repeating series of events that prepares both an ovum and the uterus for a possible pregnancy. The ovarian cycle begins when the pituitary gland secretes a hormone called follicle-stimulating hormone (FSH). This stimulates follicle development and causes the primary oocyte to restart meiosis and become a secondary oocyte. The mature follicle secretes estrogen. When the levels of estrogen reach a threshold, the hypothalamus signals the pituitary to stop producing FSH. Ovulation occurs as the follicle ruptures and releases the secondary oocyte.

When ovulation occurs, the pituitary gland produces a large amount of luteinizing hormone (LH). This hormone triggers the development of the follicle into the corpus luteum, which produces the hormone progesterone. Rising progesterone stimulates the hypothalamus to signal the pituitary to stop making LH. The corpus luteum shrinks and the levels of progesterone fall. The entire ovarian cycle restarts when the low estrogen stimulates the hypothalamus to secrete gonadotropin-releasing hormone, which restarts production of FSH by the pituitary.

The hormones released during the ovarian cycle also regulate the uterine cycle. When estrogen and progesterone are low, the endometrium disintegrates and sheds during menstruation. When estrogen levels rise again, the uterus goes into the proliferative phase and the endometrium regenerates. When the corpus luteum secretes progesterone, the uterus enters the secretory phase, and the endometrium thickens and develops mucus capable of capturing a fertilized egg.

Changes in Hormones and Events During the Human Menstrual Cycle

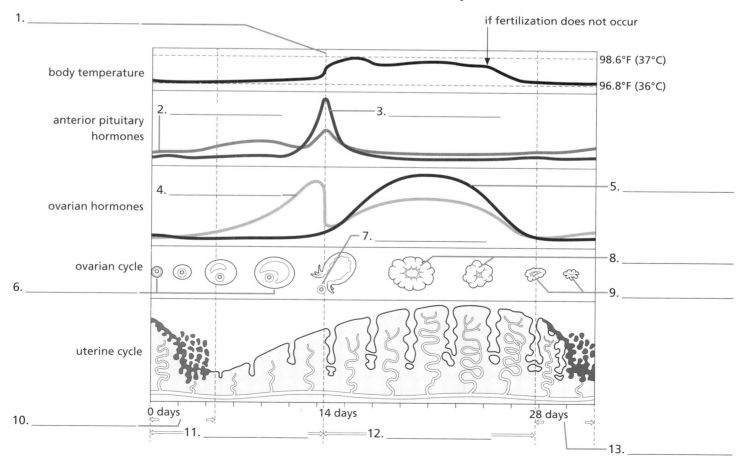

1. _____
if fertilization does not occur

body temperature — 98.6°F (37°C) / 96.8°F (36°C)

anterior pituitary hormones
2. _____   3. _____

ovarian hormones
4. _____   5. _____

ovarian cycle
7. _____
8. _____
6. _____   9. _____

uterine cycle

0 days   14 days   28 days
10. _____
11. _____   12. _____
13. _____

Embryogenesis and Fetal Development

Embryogenesis refers to the events that occur during the first eight weeks of a pregnancy. After fertilization, the zygote divides by mitosis to become a blastocyst. Pregnancy officially begins when the blastocyst implants in the endometrium and begins to produce human chorionic gonadotropin (hCG), the hormone that's detected by pregnancy tests. The endometrium thickens and develops into the placenta. The chorion and amnion from the egg develop into the amniotic sac, and the developing human reaches the embryonic stage. The amniotic sac fills with amniotic fluid.

All the organs form during the embryonic stage and are complete by about ten weeks after fertilization. The heart and major blood vessels develop, and the heart starts pumping fluid about 20 days after fertilization. The brain and spinal cord begin to grow and will continue to develop throughout pregnancy. The limbs, fingers, and toes begin to form. Because so many major developmental events occur, embryogenesis is the time when the embryo is most vulnerable to damage from teratogens such as drugs, alcohol, viruses, and radiation.

The embryo becomes a fetus around the ninth week after fertilization. All the organ systems are formed, and the fetus begins to grow. The baby's movements become more vigorous and noticeable by the mother. The sex of the baby can be determined. Toward the end of the pregnancy, fat is deposited under the skin of the fetus. The fetus rotates and drops lower in the uterus just prior to birth.

Events During Human Embryogenesis and Fetal Development and Susceptibility to Teratogens

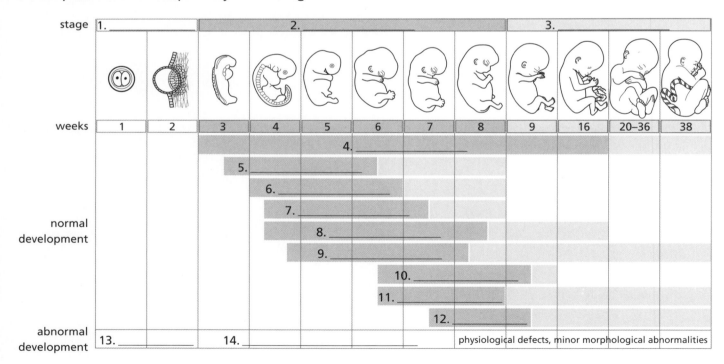

Lymphatic System

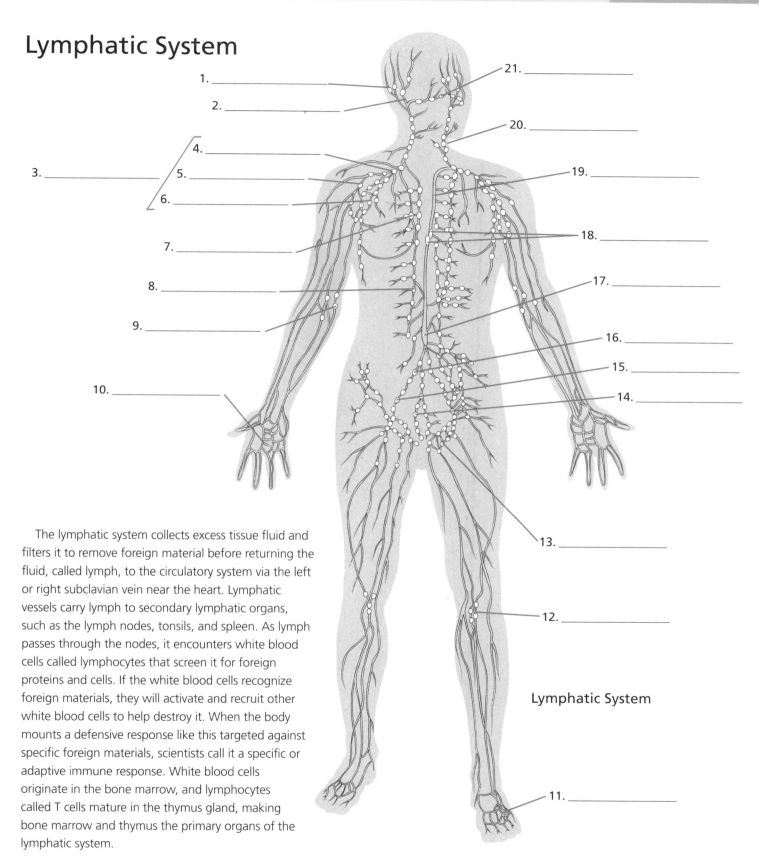

1. _____
2. _____
3. _____
4. _____
5. _____
6. _____
7. _____
8. _____
9. _____
10. _____
11. _____
12. _____
13. _____
14. _____
15. _____
16. _____
17. _____
18. _____
19. _____
20. _____
21. _____

Lymphatic System

The lymphatic system collects excess tissue fluid and filters it to remove foreign material before returning the fluid, called lymph, to the circulatory system via the left or right subclavian vein near the heart. Lymphatic vessels carry lymph to secondary lymphatic organs, such as the lymph nodes, tonsils, and spleen. As lymph passes through the nodes, it encounters white blood cells called lymphocytes that screen it for foreign proteins and cells. If the white blood cells recognize foreign materials, they will activate and recruit other white blood cells to help destroy it. When the body mounts a defensive response like this targeted against specific foreign materials, scientists call it a specific or adaptive immune response. White blood cells originate in the bone marrow, and lymphocytes called T cells mature in the thymus gland, making bone marrow and thymus the primary organs of the lymphatic system.

Answers

1. retroauricular nodes, 2. parotid nodes, 3. axillary nodes, 4. apical axillary nodes, 5. lateral group, 6. anterior group, 7. parasternal nodes, 8. posterior intercostal nodes, 9. cubital nodes, 10. palmar and dorsal plexus, 11. plantar and dorsal plexus, 12. popliteal nodes (posterior side of the knee), 13. inguinal and femoral nodes, 14. internal iliac nodes, 15. external iliac nodes, 16. common iliac nodes, 17. cisterna chyli, 18. posterior mediastinal nodes, 19. thoracic duct, 20. cervical nodes, 21. buccal nodes

Antigens and Antibodies

Antigens are molecules such as proteins, lipids, and carbohydrates that are large enough to be recognized by the immune system and trigger antibody production. Antigens that trigger an immune response include carbohydrates and lipids found in the bacterial cell wall and proteins that make up bacterial flagella and pili.

Antibodies are defensive proteins made by cells of the immune system, the lymphocytes. They have highly specific antigen-binding sites that attach to antigens, making it easier for the body to eliminate them. Each bivalent antibody is made of four polypeptide chains—two long chains and two short chains—held together by covalent bonds. The chains fold to create two antigen-binding sites per antibody. Antibodies are part of the humoral branch of the immune system because they travel through blood and other body fluids.

Individual antigens can trigger the production of multiple groups of antibodies, each of which recognizes and attaches to a specific part of the antigen called an epitope or antigenic determinant. During an immune response against a pathogen, the body produces a diverse group of polyclonal antibodies that recognize multiple epitopes. For research purposes, scientists often work with purified monoclonal antibodies that only recognize one epitope.

Structure of Antigens and Antibodies

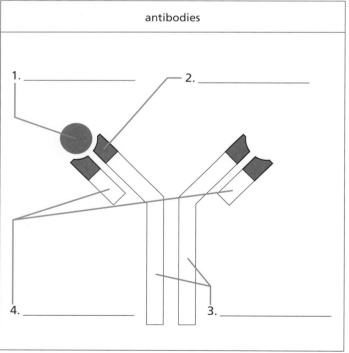

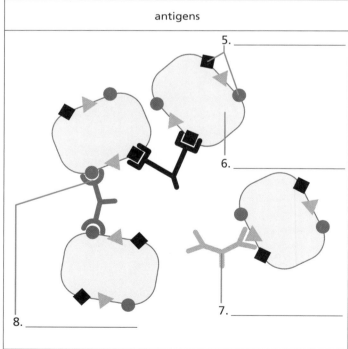

Answers

Clonal Selection and Expansion

Humans are born with a diverse set of lymphocytes that are capable of recognizing and responding to a wide variety of antigens. Until they encounter an antigen they recognize, these cells remain inactive as naive lymphocytes. When the immune system recognizes an antigen as a threat to the body, a combination of signals activates the particular set of lymphocytes that can bind to the epitopes on that antigen. The process of identifying and activating the right lymphocytes to target a particular antigen is called clonal selection. The activated lymphocytes multiply to produce clones of themselves, a process that is called clonal expansion. As cells expand into clones, cell differentiation may also occur to produce specialized cells. When activated B cells expand, for example, most cells differentiate into plasma cells, which are specialized for antibody production, while some become memory cells that enable a fast response to the same antigen in the future.

The human immune response is very powerful, and if too many cells activate at once, the response can potentially lead to negative outcomes like shock. However, humans need a large, diverse population of lymphocytes so that they can respond to the vast array of antigens they may encounter from potential pathogens. The combined processes of clonal selection and expansion balance these two conflicting needs by ensuring that the immune system first activates only the correct lymphocytes to respond to a particular antigen and then multiplies only these cells.

Response of Single Immune Cell to Stimulation by an Antigen

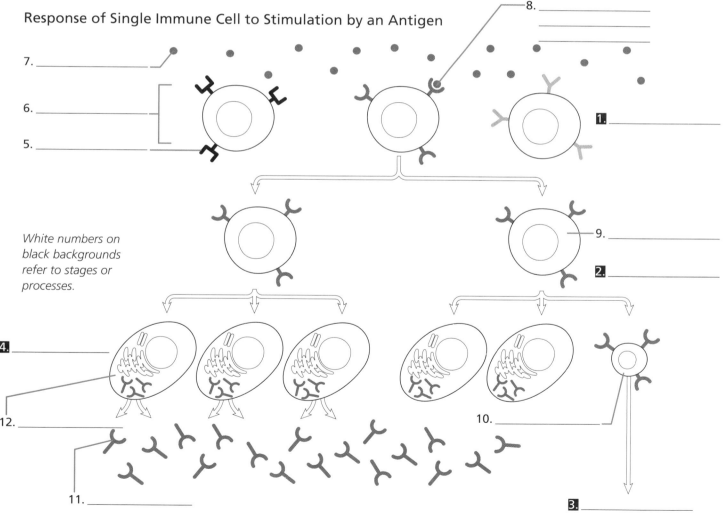

7. _____

6. _____

5. _____

8. _____
_____
_____

1. _____

9. _____

2. _____

White numbers on black backgrounds refer to stages or processes.

4. _____

12. _____

10. _____

11. _____

3. _____

Cellular Immunity: Close-Up of Helper (CD4) T Cells

Helper T cells, also known as CD4 T cells, play a central role in the adaptive immune response. Once they activate, they in turn activate cells necessary to both branches of immunity: humoral immunity, which deals with extracellular antigens circulating in the body through blood or lymph, and cellular immunity, which deals with intracellular antigens.

Helper T cells can bind only to an antigen that's presented to them by antigen-presenting cells. Antigen-presenting cells process an antigen and present it on their surfaces in a protein complex called the type II major histocompatibility complex (MHCII). Helper T cells bind to this complex with their T cell receptor. When helper T cells bind to an antigen-MHCII complex and receive other signals from immune system cells, they will activate and undergo clonal selection and expansion.

Naive helper T cells can be activated only by antigen-presenting cells called dendritic cells. If the dendritic cell has been activated by encountering a molecule it recognizes as potentially dangerous, surface proteins on the dendritic cell will bind to surface proteins on the helper T cell. This co-stimulatory signal, along with recognition and binding to the MCHII-antigen complex, activates the naive helper T cell. In response to chemical messages from itself and other immune cells, the helper T clones differentiate into helper T type I and type II cells. Type I cells activate cellular immunity, while type II cells activate humoral immunity.

Activation of a Helper (CD4) T Cell

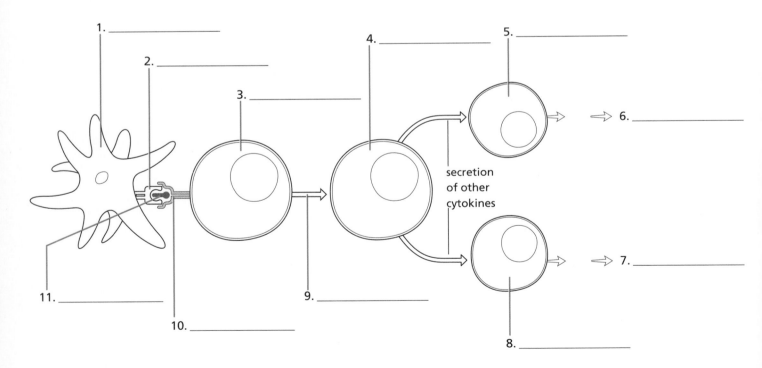

1. _____
2. _____
3. _____
4. _____
5. _____
6. _____
7. _____
8. _____
9. _____
10. _____
11. _____

secretion of other cytokines

Answers

Humoral Immunity: Close-Up of B cells

B cells are integral to humoral immunity, the branch of immunity that deals with extracellular antigens circulating in the body through blood or lymph. B cells activate when they bind antigen with their B cell receptors. Once activated, they have the potential to produce clones that differentiate into plasma cells and memory B cells. Plasma cells produce antibodies, while memory B cells remain in an activated state so they can respond quickly if they re-encounter the antigen they recognize. Memory B cells are an important component of the immunologic memory that allows humans to resist diseases they've either experienced before or been vaccinated against.

T-dependent activation is required for full differentiation of B cell clones into plasma cells and memory B cells. During this interaction, an activated B cell processes and presents an antigen on its surface in the type II major histocompatibility (MHCII) complex. An activated helper T cell binds the antigen-MHCII complex with its T cell receptor. Additionally, surface proteins on the helper T cell bind to surface proteins on the B cell, providing an important co-stimulatory signal to the B cell. The combination of signals from independent recognition of an antigen and the co-stimulatory signals from the helper T cells triggers expansion and differentiation of B cell clones.

How Helper T Cells Activate B Cells

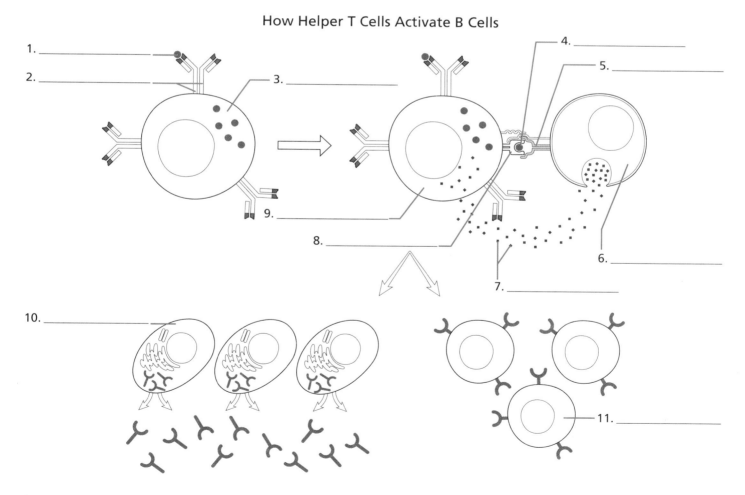

Cellular Immunity: Close-Up of Cytotoxic (CD8) T cells

Cytotoxic T cells, also called CD8 T cells, play a critical role in cellular immunity. Like helper T cells, they can be activated only by antigen that's presented on the surface of antigen-presenting cells. Cytotoxic T cells, however, bind to an antigen that's presented in a protein complex called type I major histocompatibility complex (MHCI). This protein complex is found on the surface of all nucleated cells.

Naive cytotoxic T cells activate when they detect antigen presented in MHCI on the surface of dendritic cells. If the dendritic cell is active, its surface proteins will bind to surface proteins on the cytotoxic T cell. The combination of these two signals will trigger clonal expansion in the cytotoxic T cell.

Activated cytotoxic T cells scan the surface of body cells by attempting to bind to their MHCI complexes with their T cell receptors. If the body cell is infected with a pathogen such as a virus, it will process any intracellular foreign antigen and display it in its MHCI complexes.

If an activated cytotoxic T cell is able to bind to the MHCI-antigen complex, it will release proteins called perforins and granzymes to trigger programmed cell death, or apoptosis, in the infected cell. Perforins are straw-shaped proteins that perforate the plasma membrane of the infected cell. Granzymes are enzymes that enter the cell and activate apoptosis. The infected cell dies, preventing the pathogen from replicating further.

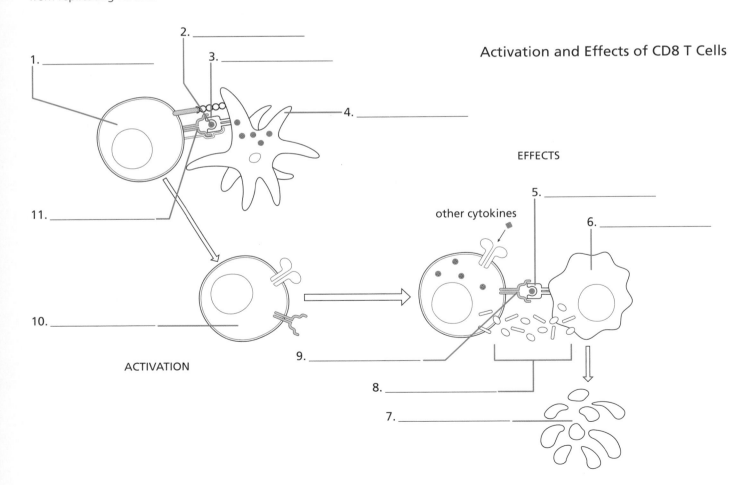

Activation and Effects of CD8 T Cells

1. _____

2. _____

3. _____

4. _____

EFFECTS

5. _____

other cytokines

6. _____

11. _____

10. _____

ACTIVATION

9. _____

8. _____

7. _____

Latitude and Solar Radiation

Biogeography is the study of the global distribution of organisms. The number and diversity of organisms found in any area depends on a combination of living (biotic) and nonliving (abiotic) factors. The particular range, or geographic distribution, of any one species depends on its ability to tolerate the climate of the area, to find food, to avoid predation, and to disperse its offspring.

Climate refers to the typical long-term weather conditions of an area. Many factors, including latitude, wind patterns, and geographic features such as mountains, combine to determine the climate of an area. Latitude is important because it determines the amount of solar radiation an area receives. In general, areas of the world that receive large amounts of sunlight, such as those near the equator, are warmer than those that receive less sunlight, such as those at the poles.

The curvature of the Earth affects the amount of solar radiation an area receives. At the equator, the sun is often directly overhead, so that its rays strike the Earth parallel to the latitude lines, resulting in a maximum amount of solar radiation per unit area. The surface of the Earth curves away from the equator, so that incoming radiation strikes at lower and lower angles moving toward the poles. The radiation is therefore more diffuse, and the total energy per unit area is less.

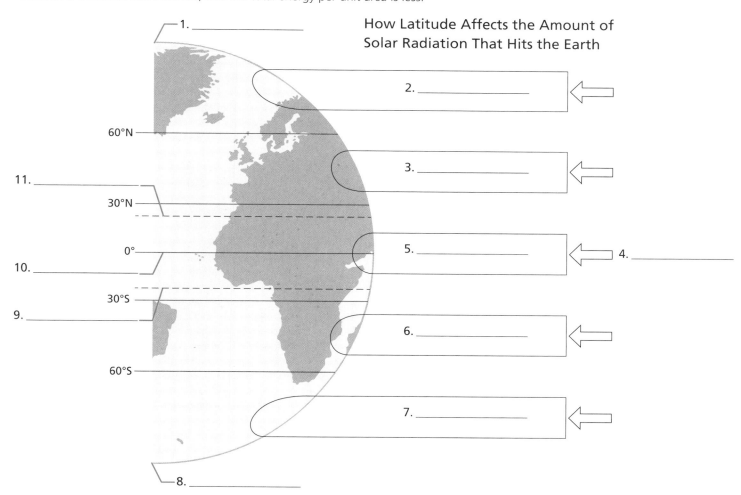

How Latitude Affects the Amount of Solar Radiation That Hits the Earth

Answers

Global Air Circulation

The air in the Earth's atmosphere is constantly moving due to changes in temperature and pressure. As the Earth orbits the sun and rotates on its tilted axis, certain parts of the atmosphere receive more solar radiation than others. The gases expand as air warms, making the air less dense and causing it to rise, which creates a low-pressure zone at the Earth's surface. The warm air spreads out in the upper atmosphere, where it cools and increases in density. The cooler air sinks again, and where it meets the surface it moves toward the low-pressure zone, completing a circuit that scientists call a convection cell. The movement of air within a convection cell twists slightly due to the Coriolis effect, which is caused by the rotation of the Earth.

Because warm air holds more moisture than cool air, convection cells affect precipitation. As warm air rises, it carries moisture up into the atmosphere. As the air cools, it can't hold as much water and it releases precipitation on the land below. As the cool, dry air begins to sink again, it warms as it approaches the Earth's surface, gaining capacity to hold moisture. The warm air absorbs water from the surrounding area as it travels over it, making the land very dry.

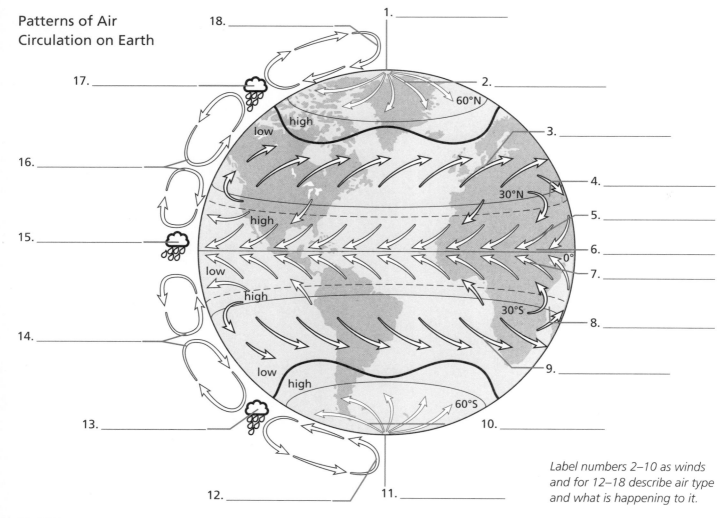

Patterns of Air
Circulation on Earth

1. _____
2. _____
3. _____
4. _____
5. _____
6. _____
7. _____
8. _____
9. _____
10. _____
11. _____
12. _____
13. _____
14. _____
15. _____
16. _____
17. _____
18. _____

60°N
30°N
0°
30°S
60°S
high
low
high
low
high
low
high

Label numbers 2–10 as winds and for 12–18 describe air type and what is happening to it.

Answers

1. North Pole, 2. polar easterlies, 3. westerlies, 4. horse latitudes, 5. northeast trade winds, 6. doldrums, 7. southeast trade winds, 8. horse latitudes, 9. westerlies, 10. polar easterlies, 11. South Pole, 12. descending cold, dry air, 13. rising warm, moist air, 14. descending cold, dry air, 15. rising warm, moist air, 16. descending cold, dry air, 17. rising warm, moist air, 18. descending cold, dry air

The Rain Shadow Effect

Landmasses near oceans often have milder climates due to the ability of water to store heat energy. Oceans absorb heat from the atmosphere during the summer, helping keep nearby landmasses cooler. Then, during the winter, the ocean releases heat to the atmosphere, keeping coastal regions warmer than those inland.

When air travels over mountains, the changes in elevation can affect precipitation on either side of the mountain. As warm, moist air passes over a mountain range, it cools as it reaches higher elevations. As it cools, it loses its ability to hold moisture and releases precipitation onto the land below. The cool, dry air descends on the other side of the mountain range, warming as it reaches lower elevations. This warm, dry air absorbs moisture from the surrounding area, creating desert conditions. This drying effect due to a mountain range is called a rain shadow.

How a Rain Shadow Affects Precipitation

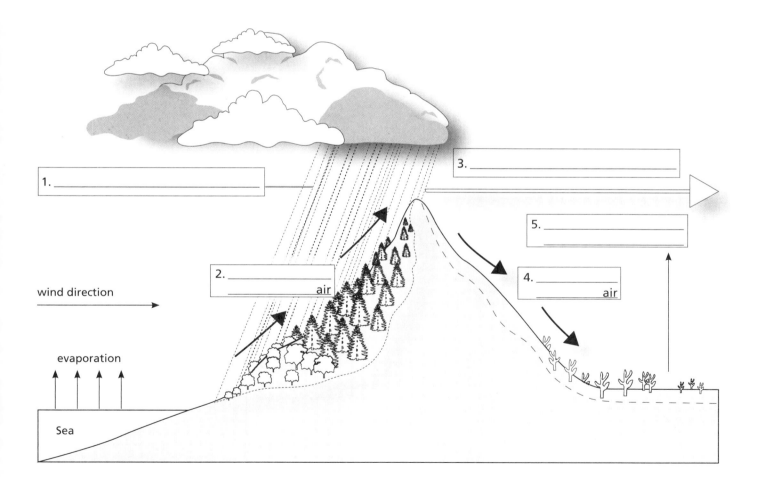

Answers

1. rain from compression and cooling, 2. warm, moist, 3. region of rain shadow, 4. cool, dry, 5. evaporation from expansion and warming

Ocean Currents

The pattern of ocean currents around the world depends on factors such as wind, water density, and tides. Local geographic features, such as coastlines and underwater mountain ranges, also have an effect. As ocean currents flow, they twist slightly because of the Coriolis effect, creating circular ocean currents called gyres. Gyres in the northern hemisphere twist clockwise, while those in the southern hemisphere twist counterclockwise.

Differences in density cause ocean waters to rise and fall, generating deep ocean currents that connect to surface currents and move water around the planet. When ice forms in polar oceans, it removes pure water from solution, leaving the remaining fluid water with a higher salt concentration. This cold, salty water sinks because it is more dense than the water below it. Water flows in to replace the saltier water, creating a cycle of convection that drives the deep ocean currents.

Global winds create surface currents by pushing the water as they flow past. These surface currents have large impacts on local climates. For example, the Gulf Stream brings warm water from the tropical regions of the Caribbean to much of northern Europe, resulting in a milder climate there than in other northern regions. In contrast, Peru is cooler than many of the countries that surround it because the offshore Peru or Humboldt Current brings cooler water northward along the South American coast.

Pattern of Ocean Currents on Earth

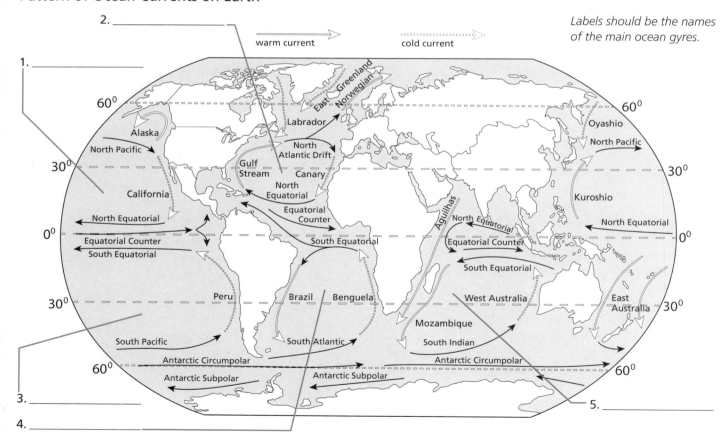

Labels should be the names of the main ocean gyres.

Answers

Terrestrial Biomes

The average annual precipitation and temperature of a region have a huge impact on determining which organisms can live there. Scientists have identified approximately 10 major groups of distinct plants and animals that form ecological communities, called biomes, on planet Earth. Biomes can be subdivided into ecosystems, which include all the organisms in an area interacting with each other and with the abiotic components of their environment. Many varieties of ecosystems can exist within one biome.

Terrestrial biomes include tropical rainforests, deserts, temperate grasslands, temperate forests, boreal forests, and the arctic tundra. Each of these biomes has a unique profile of annual average temperature and precipitation. When warm temperatures combine with high rainfall, as they do in the tropical rainforest, the result is one of the most productive and diverse biomes on the planet. High temperatures without rain, however, result in low productivity because of slow plant growth, as in deserts. Biomes in temperate areas strike a middle ground, with moderate rainfall and temperatures that permit plant growth for most of the year. Cold temperatures limit the growing season, leading to biomes with lower productivity like boreal forests and the arctic tundra. In these areas, the average temperatures are below freezing for much of the year, and much of the soil is permanently frozen (permafrost).

KEY

1. _____
2. _____
3. _____
4. _____
5. _____
6. _____
7. _____
8. _____
9. _____
10. _____

The Various Terrestrial Biomes on Earth

Atlantic Ocean

Pacific Ocean

Indian Ocean

Answers

1. tundra, 2. boreal forest, 3. temperate grassland, 4. desert, 5. chaparral, 6. savanna, 7. temperate broadleaf forest, 8. tropical rainforest, 9. high mountains, 10. polar ice

Zones in Oceans and Lakes

Salinity and water flow, as well as the availability of light and nutrients, create distinct aquatic biomes such as lakes, streams, freshwater wetlands, estuaries, and oceans. Wetlands are shallow-water habitats where the soil is saturated with water for at least part of the year. Estuaries are areas where fresh water and salt water combine. In lakes and oceans, the amount of available light creates unique zones that are inhabited by different types of organisms. The area penetrated by light is called the photic zone; areas that don't receive light are in the aphotic zone.

The greatest abundance of life in the ocean occurs along the continental shelf, which is the relatively shallow area close to shore. The intertidal zone, which is closest to shore, receives enough sunlight to support photosynthesis but has the challenge of being exposed to the air as the tide goes in and out. The neritic zone extends out from the intertidal zone to the edge of the continental shelf, which is where the oceanic zone begins. Scientists call the bottom of the ocean the benthic zone.

Like oceans, lakes can be divided into zones such as the photic, aphotic, and benthic zones. The area along the shore of a lake that is shallow enough for rooted plant growth is called the littoral zone. The area of the lake that's too deep for rooted plants but that does receive light is called the limnetic zone.

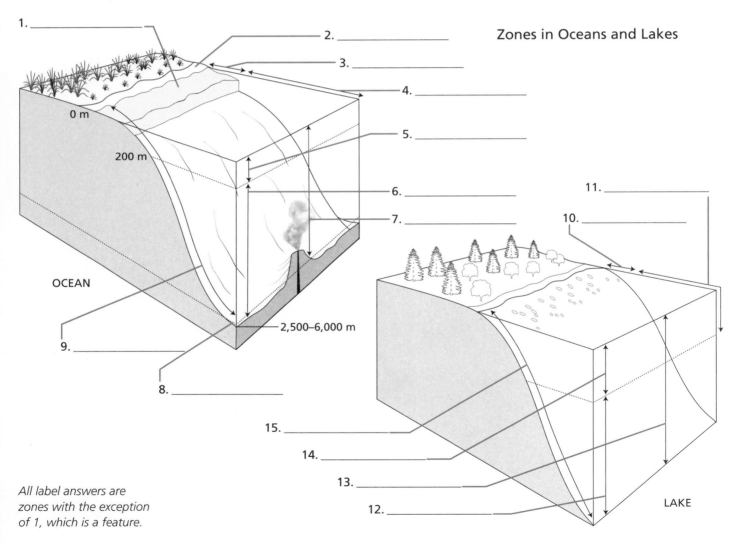

Zones in Oceans and Lakes

1. _____
2. _____
3. _____
4. _____
5. _____
6. _____
7. _____
8. _____
9. _____

0 m
200 m
2,500–6,000 m
OCEAN

10. _____
11. _____
12. _____
13. _____
14. _____
15. _____

LAKE

All label answers are zones with the exception of 1, which is a feature.

Answers

Upwelling and Turnover

Upwelling and turnover arhye mechanisms that increase nutrient availability in oceans and lakes. Because nutrient availability affects the growth of photosynthetic organisms, it also affects the numbers of organisms that can be supported in these ecosystems.

Upwelling occurs in oceans when winds push surface waters away from the shore, causing cold water to rise up from below. The water brings with it nutrients from the ocean floor, which support the growth of photosynthetic organisms such as phytoplankton. Phytoplankton are the basis of food webs that support larger organisms such as fish. Typically, good fishing grounds occur where upwelling is common.

Turnover in lakes occurs because of differences in seasonal temperatures. During the summer, the surface water is warmer and less dense than the water below it. The water of the lake becomes stratified into layers, with the warm epilimnion on top and a cold hypolimnion below. Scientists call temperature gradients like this a thermocline. In the fall, the water in the epilimnion cools and becomes more dense, sinking and displacing the nutrient-rich water in the hypolimnion below. In the transition from winter to spring, turnover occurs again in response to density and temperature changes.

Processes of Upwelling and Turnover

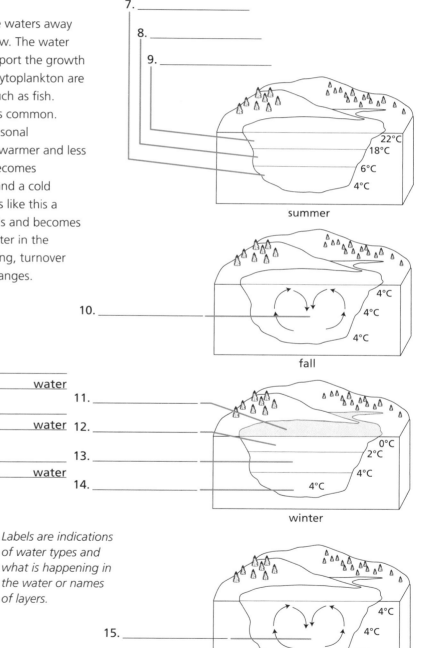

7. _____

8. _____

9. _____

22°C
18°C
6°C
4°C

summer

10. _____

4°C
4°C
4°C

fall

11. _____
12. _____
13. _____
14. _____

0°C
2°C
4°C
4°C

winter

15. _____

4°C
4°C
4°C

spring

longshore wind

1. _____ water

100 m
200 m   2. _____ water

300 m
400 m   3. _____ water

4. _____ water

Equator   Northern Hemisphere
Southern
Hemisphere

5. _____

6. _____ water

Labels are indications of water types and what is happening in the water or names of layers.

Answers

Energy Flow in Ecosystems

Many different types of interactions can occur within an ecosystem, such as competition for resources, mates, or habitat space. One of the most common purposes of these interactions is to obtain energy. Primary producers can capture energy from abiotic sources such as the sun. Herbivores get energy by eating producers. Consumers eat herbivores or other consumers. Decomposers eat all types of organisms once they die.

Ecologists say that energy flows through ecosystems. Ultimately, the source of all the energy for almost all life on Earth is the sun. Photosynthetic organisms, such as plants, absorb this sunlight and use it to make carbohydrates from carbon dioxide and water, ultimately converting light energy to chemical energy. Every other living thing eats the plants or eats something that eats the plants. In the process of transferring energy from food to ATP, usually via cellular respiration, all living things lose some of that energy to the environment as heat. So, energy flows into ecosystems from the sun, gets stored in plants, is transferred to other organisms, and eventually ends up as heat in the atmosphere. The heat in the atmosphere is lost back to space.

How Energy Moves Through Ecosystems

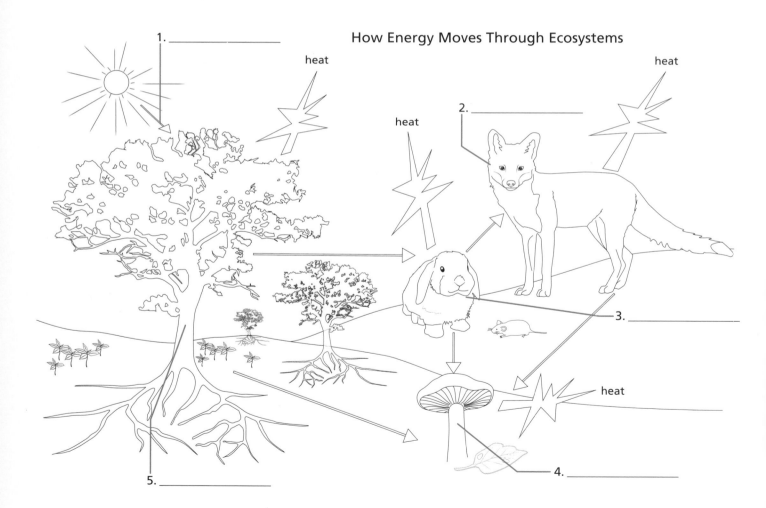

1. _____

heat

heat

2. _____

heat

3. _____

4. _____

5. _____

Answers

1. light energy, 2. secondary consumer (carnivore), 3. primary consumer (herbivore), 4. decomposer, 5. producer

Food Webs

Food is a necessary source of matter and energy for all organisms. (Even photosynthetic organisms need food; they just make their own.) Food chains link organisms together to show how food moves from one organism to the next through predation. For example, a food chain could connect a leaf, a caterpillar, and a bird. The leaf is a producer. The caterpillar is a primary consumer because it feeds directly on the leaf. The bird is a secondary consumer because it eats a primary consumer. Scientists draw arrows to connect these organisms in a chain, with the arrowhead pointing in the direction of the flow of matter and energy.

Most organisms in an ecosystem eat more than one type of food. When diverse food sources are included, food chains become interwoven to create a food web. Food webs help illustrate the interconnections between the organisms in an ecosystem. This information can be used to understand problems or predict impacts of species loss. For example, if a species is threatened, then organisms that eat that species may also be negatively impacted. The potential impact depends on the number of connections within the web. Generalists that eat many different things are less likely to be impacted, while specialists with limited food sources are typically more at risk for extinction if their food sources are reduced.

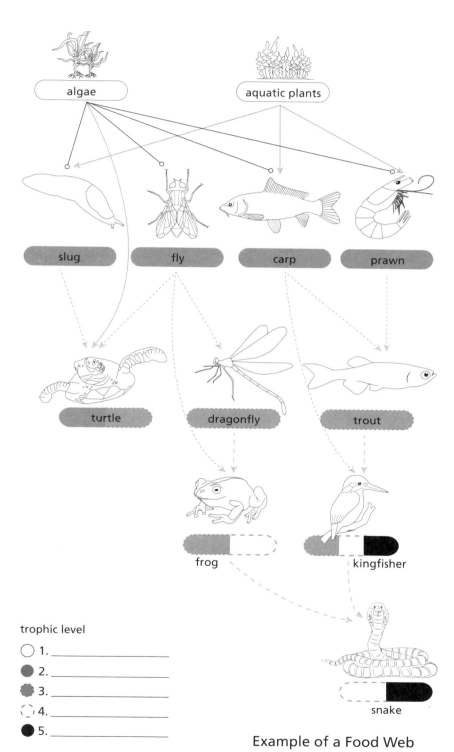

Example of a Food Web

trophic level

○ 1. _____

● 2. _____

● 3. _____

◌ 4. _____

● 5. _____

Answers

1. producers, 2. primary consumers, 3. secondary consumers, 4. tertiary consumers, 5. quaternary consumers

The Energy Pyramid

Energy pyramids are another way to look at the flow of energy through ecosystems. The pyramid is organized by stacking the categories of organisms on top of each other, with producers at the bottom. Each layer of the pyramid can also be referred to as a trophic level (troph- means "food"). The pyramid shape provides a visual cue for what happens to the usable energy in an ecosystem as energy flows from one category to the next. The first trophic level, represented by the producers, has the most stored energy and therefore is the largest layer, while the trophic level at the top of the pyramid has the least and is the smallest layer.

Plants transfer about 1 percent of the available light energy to chemical energy in carbohydrates and then use this stored energy to grow and reproduce. With every energy transfer in metabolism, some available energy becomes heat in the environment. Because producers use much of their stored energy, and lose some of it to the environment as heat, only about 10 percent of the energy they originally stored in food is available to the primary consumers in the second trophic level. Similarly, the primary consumers use most of the energy they capture by grazing on producers for their own growth and reproduction. In general, about 10 percent of the energy captured by a trophic level is available to pass to the next level. Ecologists call this the "10 percent rule."

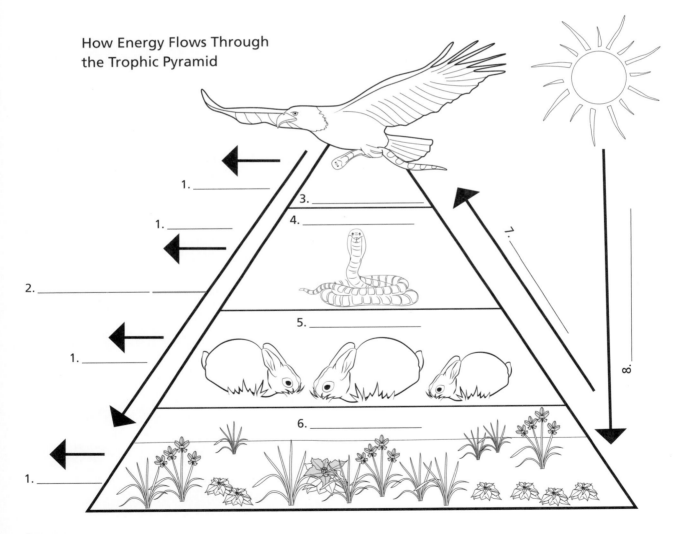

How Energy Flows Through the Trophic Pyramid

1. _____
1. _____
2. _____
1. _____
1. _____
3. _____
4. _____
5. _____
6. _____
7. _____
8. _____

Answers

Nutrient Cycles

Ecologists say that matter cycles within ecosystems. Almost all the matter that's here on planet Earth has been here since the planet first formed about 4.6 billion years ago. Atoms are incorporated into molecules, molecules change phases, and living things are born and then die, but the atoms don't disappear—they are constantly recycled and reused. Using a propane (C_3H_8) grill is a familiar example. When you make a spark in the presence of oxygen (O_2), the propane burns. Even though the level in the tank gets lower, the atoms don't disappear; instead, they change bonding partners to become carbon dioxide and water vapor in the atmosphere.

Fixation or assimilation refers to processes that capture atoms from the environment and incorporate them into molecules in living things. These atoms return to the environment through the processes of digestion, excretion, and decomposition. Under certain conditions, the remains of living things undergo fossilization to become fossils or fossil fuels. Combustion of fossil fuels returns atoms to the atmosphere.

Atoms also change forms due to physical processes. Minerals in sediments can be compressed into sedimentary rocks. Rocks undergo transformations through conditions of high pressure or heat within the Earth. Minerals contained in rocks may be unavailable for assimilation by living things. However, exposure to wind, ice, and water over time causes weathering and erosion of these rocks, returning minerals to small sediments that are again available to life.

Overview of How Nutrients Cycle in Ecosystems

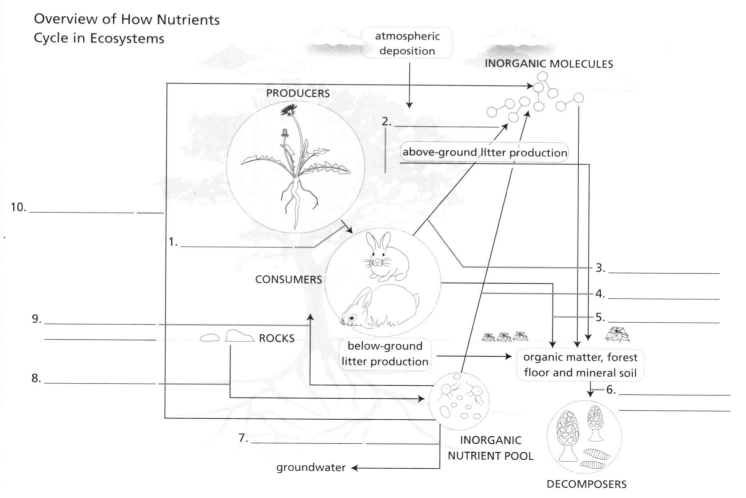

Answers

1. predation, 2. photosynthesis, 3. respiration, 4. fixation, 5. death, 6. decomposition and mineralization, 7. leaching, 8. mineral weathering, 9. nutrient uptake and assimilation, 10. gaseous loss

The Water Cycle

The water cycle explains how water moves around the Earth between the oceans, land, and atmosphere. As water moves through this cycle, it undergoes phase changes between its solid, liquid, and gas forms. Because organisms require available water, the water cycle has a big impact on ecosystems.

The oceans contain most of the water on Earth. Glaciers and permanent snow store small amounts, as do bodies of water such as lakes and streams. Only a tiny amount occurs in the atmosphere as water vapor, but this tiny amount has a big impact on the Earth's weather. Water vapor is a greenhouse gas that holds heat and deflects it back down to the Earth's surface.

Precipitation, such as snow or rain, falls to the Earth as water vapor in the atmosphere condenses. Precipitation may fall directly into bodies of water, or it may be stored temporarily in snow. As snow melts, the water runs into streams and is eventually carried to the oceans. Water can also move through porous stone as groundwater before making its way to the oceans. Energy from solar radiation causes liquid water to transform back to water vapor, returning water to the atmosphere. Organisms can also release water vapor from processes such as transpiration and cellular respiration.

Steps in the Water Cycle

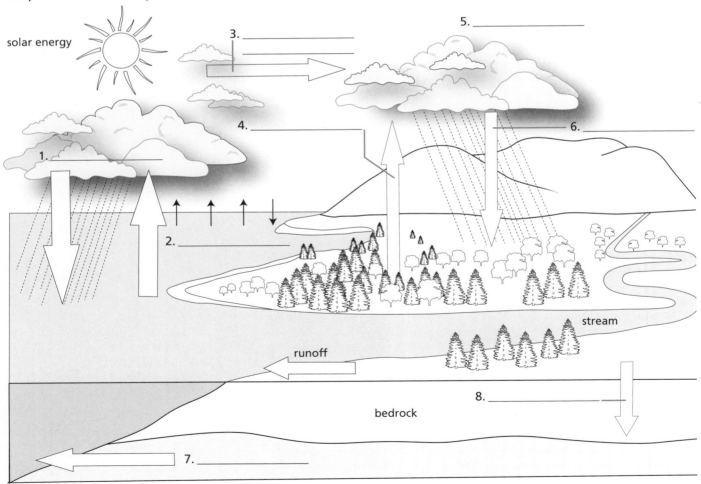

Answers

1. precipitation over the sea, 2. evaporation from the sea, 3. transport over the land, 4. evapotranspiration, 5. movement of water vapor by wind, 6. precipitation over the land, 7. groundwater runs to the sea, 8. infiltration

The Carbon Cycle

The carbon cycle describes the way that carbon moves through ecosystems. Carbon exists in the atmosphere and oceans as carbon dioxide. Photosynthetic organisms capture carbon dioxide and convert it to organic molecules such as sugars. These organic molecules carry some carbon through the trophic levels of food webs as organisms eat other organisms. When organisms die, the carbon in their bodies briefly becomes part of the organic matter in the soil until their remains are eaten by decomposers. As decomposers and other organisms use cellular respiration to get energy from food molecules, they release carbon back to the atmosphere as carbon dioxide. Combustion of wood and fossil fuels also releases carbon dioxide to the atmosphere.

Humans have had a tremendous impact on the carbon cycle over the past 300 years. The Industrial Revolution fueled the development of human cities with the burning of fossil fuels such as coal and oil. These carbon-rich fuels developed from the remains of organisms deposited in the Earth during the Carboniferous period (359–299 million years ago). By burning large quantities of these fuels, humans increased the concentration of carbon dioxide in the atmosphere. Because carbon dioxide is a greenhouse gas, global temperatures have begun to rise. The warming temperatures affect the water cycle and global weather patterns, and they also reduce the ability of the oceans to store dissolved carbon dioxide.

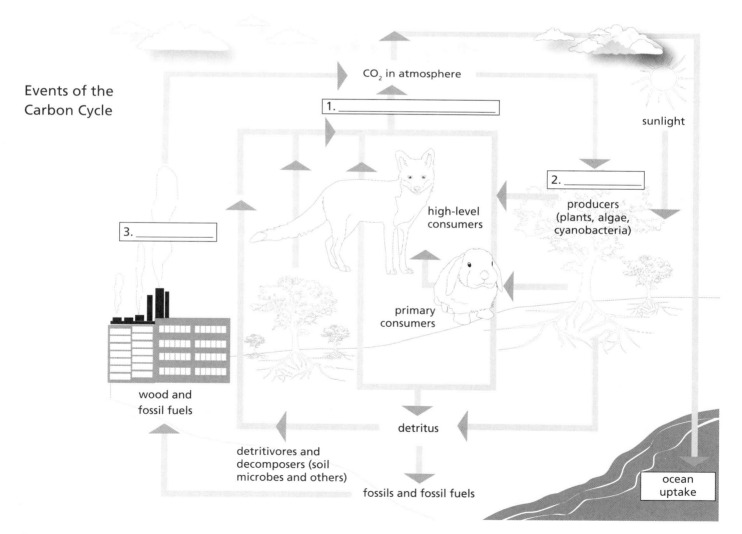

Events of the Carbon Cycle

CO_2 in atmosphere

1. _____

sunlight

2. _____

producers (plants, algae, cyanobacteria)

high-level consumers

3. _____

primary consumers

wood and fossil fuels

detritus

detritivores and decomposers (soil microbes and others)

fossils and fossil fuels

ocean uptake

Answers

The Nitrogen Cycle

The nitrogen cycle refers to the movement of nitrogen through ecosystems. The largest reservoir for nitrogen is the atmosphere, which is almost 80 percent nitrogen gas (N_2). Nitrogen-fixing bacteria capture this nitrogen, converting it into ammonia (NH_3) and using it to build their own proteins and nucleic acids. When bacteria release ammonia to the soil, scientists call it ammonification. Other bacteria, called nitrifying bacteria, use ammonia as a source of energy and produce nitrite (NO^{2-}) and nitrate (NO^{3-}) as wastes. Bacteria called denitrifying bacteria use nitrate as a substitute for oxygen during cellular respiration and produce nitrite or nitrogen gas as waste.

Plants can assimilate ammonia or nitrate from the soil, using these forms of nitrogen for their proteins and nucleic acids. The nitrogen moves through the trophic levels of the ecosystem as organisms eat each other. When organisms die, decomposers eat their remains. When decomposers break down proteins, they release excess nitrogen as ammonia into the soil.

Lightning and volcanoes also contribute to the nitrogen cycle: lightning strikes can produce nitrate and volcanoes emit some ammonia. Humans use energy to fix nitrogen into chemical fertilizers, which they use to add nitrate to the soil for plant growth. Human factories also produce some nitrogen compounds as waste from industrial processes.

Steps in the Nitrogen Cycle

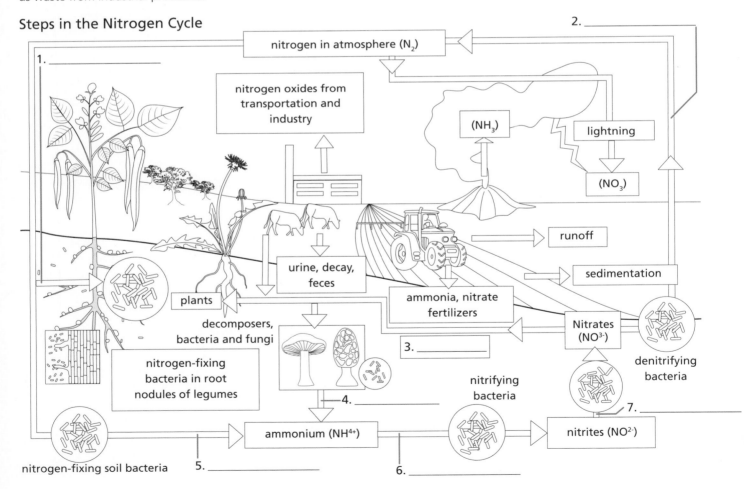

2. _____

nitrogen in atmosphere (N_2)

1. _____

nitrogen oxides from transportation and industry

(NH_3)

lightning

(NO_3)

runoff

sedimentation

urine, decay, feces

plants

ammonia, nitrate fertilizers

Nitrates (NO^{3-})

denitrifying bacteria

decomposers, bacteria and fungi

3. _____

nitrogen-fixing bacteria in root nodules of legumes

4. _____

nitrifying bacteria

7. _____

nitrogen-fixing soil bacteria

5. _____

ammonium (NH^{4+})

6. _____

nitrites (NO^{2-})

Answers

1. nitrogen fixation, 2. denitrification, 3. assimilation, 4. ammonification, 5. nitrogen fixation, 6. nitrification, 7. nitrification

The Greenhouse Effect

The greenhouse effect is the warming effect that the Earth's atmosphere has on surface temperatures. Just like the glass of a greenhouse helps hold heat inside, our atmosphere holds heat around the planet. When light energy from the sun arrives at the Earth, some of it is absorbed and some of it reflects off the surface of the planet and bounces back toward space. The light energy that's absorbed is eventually re-emitted from the surface as heat. Without any atmosphere, the energy from the sun would end up back in outer space and the surface of the Earth would be as cold as the Moon. Greenhouse gases in the atmosphere, such as carbon dioxide and water vapor, absorb energy as it travels toward space and re-emit it back down to the surface. This retains the energy at the Earth's surface longer and warms the planet.

The greenhouse effect is necessary to sustain life as we know it. However, in recent years, human activity has started to affect its impact on Earth's temperature. Since the Industrial Revolution, humans have burned large amounts of fossil fuels, adding a great deal of carbon dioxide to the atmosphere. This has increased the greenhouse effect, causing a rise in average global temperatures, polar ice melt, and the frequency of severe weather events. Scientists predict grave outcomes for humans and other organisms if we don't lower our carbon dioxide emissions.

Factors That Contribute to the Greenhouse Effect

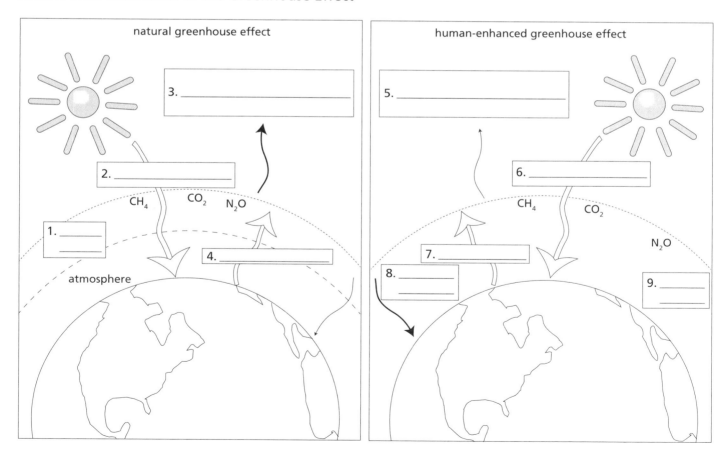

Answers

Bioaccumulation and Biomagnification

Bioaccumulation occurs when organisms accumulate toxic chemicals in their bodies. Some chemicals, such as polychlorinated biphenyls (PCBs), persist in the environment for a long time, resulting in prolonged exposure for organisms living in that environment. Organisms can absorb the chemicals through their skin or ingest them in food or water. If they take in the chemicals faster than they excrete them, the chemicals concentrate in their tissues. Chemicals that are fat-soluble are particularly problematic for animals because they accumulate in long-term fat storage. When the chemicals reach a certain threshold, they can cause a variety of effects, including negative impacts on neurological and reproductive systems.

Bioaccumulation can lead to the problem of biomagnification, which is the increasing concentration of a toxic chemical as you move through the trophic levels of an ecosystem. When an organism eats another, it can absorb any toxic chemicals the organism contains into its own tissues. Because each organism feeds on many others, the concentration of the chemical increases as it moves through the trophic pyramid. For example, mercury is present in very small amounts in seawater, but it accumulates in algae when they absorb it. Zooplankton eat the algae, and the concentration of mercury begins to magnify. Small fish eat the zooplankton, and bigger fish eat the small fish. As a result, higher-level predators like swordfish (*Xiphias gladius*) and tuna have a much greater amount of mercury in their tissues than is found in seawater itself.

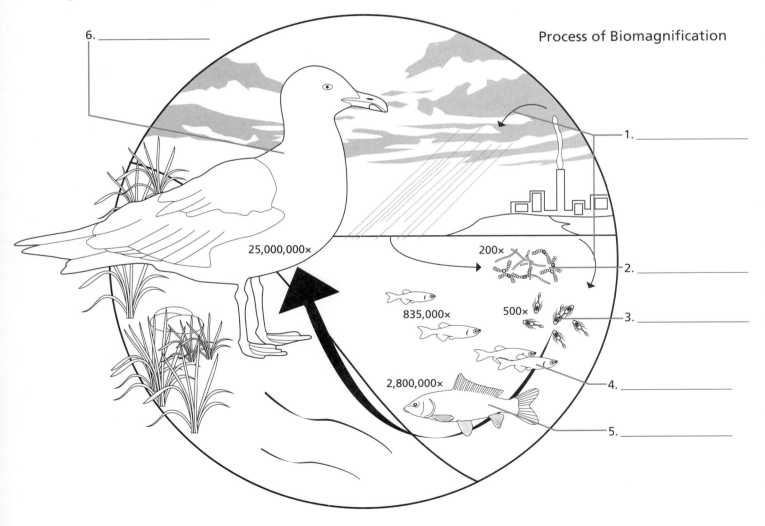

Process of Biomagnification

6. _____

1. _____

2. _____

3. _____

4. _____

5. _____

25,000,000×

200×

500×

835,000×

2,800,000×

Answers

Population Distribution

A population is the total number of a particular species that live in an area. The science of demographics looks at the structure of that population and how it changes over time. Two of the most informative measurements of a population are the population size, which is the total number of individuals that live in the area, and the population density, which is the number of individuals divided by the size of the area (number per unit of area or volume).

Ecologists also look at how a population is distributed. Populations may exhibit random distribution, in which individuals are randomly spaced, without a predictable pattern. Populations may also show uniform distribution, in which individuals are evenly spaced. Many populations exhibit clumped distribution, where some areas have higher numbers of individuals and some have fewer.

Population density gives a rough estimate of how crowded a population is, but it can be misleading depending on how the population is dispersed. If a population exhibits uniform (even) distribution, then population density is a very good representation of the amount of crowding. However, for human populations with clumped distribution, the average population density of a country would not accurately represent the density in either the cities or the rural areas.

Types of Population Distribution

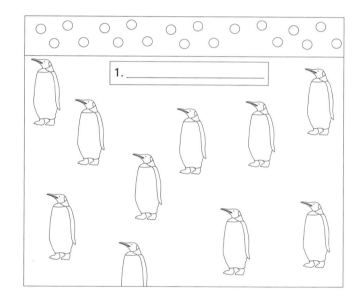

1. _____

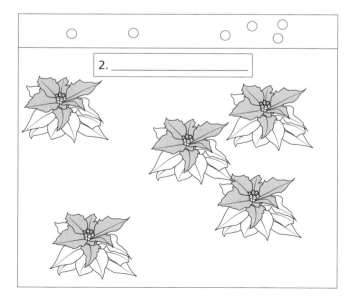

2. _____

3. _____

Answers

1. uniform, 2. random, 3. clumped

Population Growth

Population growth is determined by the number of organisms in a population (N), the rate at which they reproduce (r), and the number of organisms a particular ecosystem can support, which is called the carrying capacity (K). Different reproductive strategies may be advantageous in different conditions. For example, in rapidly changing environments, organisms that can reproduce quickly might have an advantage because they can respond rapidly to favorable conditions. A quick burst of reproduction might cause them to overshoot the carrying capacity of their environment, however, leading to a subsequent population crash. Because this strategy relies on a fast reproduction rate, ecologists say that these organisms are r-selected. In environments where the population is already close to carrying capacity, a more successful strategy might be to produce fewer offspring but invest in them to ensure their survival. Because these organisms live near the carrying capacity, ecologists say that they're K-selected.

Ecologists also look at populations in terms of how long groups, or cohorts, of organisms survive. A cohort is a group of organisms born at the same time. When ecologists map the survivorship of cohorts over time, they see three general patterns. Cohorts that have high survivorship for most of their lifespan and then rapidly die off exhibit Type I survivorship. Cohorts that have steady death rates throughout their lifespan exhibit Type II survivorship. Cohorts that have very high mortality at young ages exhibit Type III survivorship.

Different Strategies for Population Growth

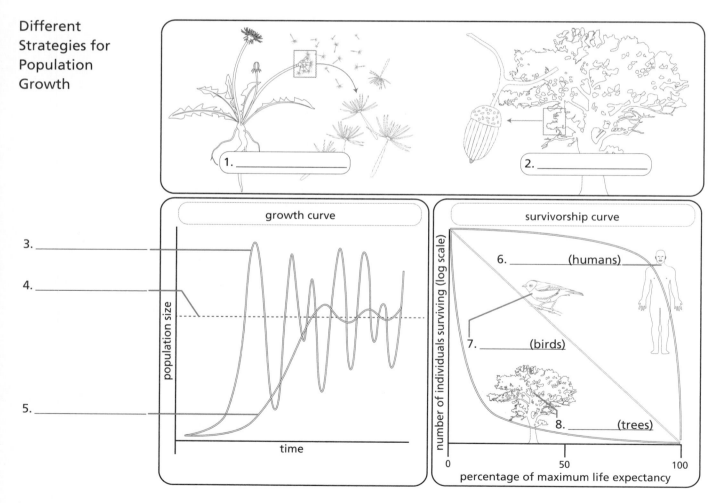

1. _____

2. _____

growth curve

3. _____

4. _____

5. _____

population size

time

survivorship curve

number of individuals surviving (log scale)

6. _____ (humans)

7. _____ (birds)

8. _____ (trees)

percentage of maximum life expectancy

0   50   100

Density-Dependent Factors

Factors That Affect Population Size

White numbers on black backgrounds are column headings.

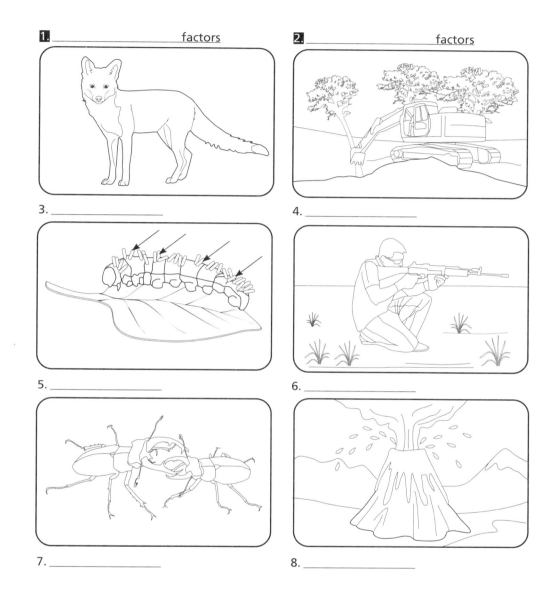

1. _____ factors

2. _____ factors

3. _____

4. _____

5. _____

6. _____

7. _____

8. _____

Density-dependent factors affect population sizes with greater or lesser impact depending on the size of the population. Infectious disease is a density-dependent factor because pathogens spread more quickly when organisms are crowded. Also, the size of some predator populations rises and falls with changes in the size of their prey population. When organisms are crowded, they must compete with each other for space, food, and other resources. Thus, competition is also a density-dependent factor that can affect population size.

The impact of density-independent factors doesn't change with population size. Habitat destruction reduces populations whether they are crowded or not. Climate change will affect all the organisms in an ecosystem regardless of their population density. Likewise, natural disasters such as fires or floods are density-independent.

Answers

1. density-dependent, 2. density-independent, 3. predation, 4. habitat destruction, 5. disease 6. hunting by humans, 7. competition, 8. natural disasters

Age Structure of Human Populations

People who study the demographics of human populations use population pyramids, or age-structure diagrams, to visualize the distribution of age groups in a population. The diagram represents the number of people in each age group as a horizontal bar. The bars are stacked on top of each other, with the youngest age group at the bottom. The shape of the diagram demonstrates whether the population is growing, shrinking, or staying the same.

    An age-structure diagram that forms a pyramid shape shows that the population is still growing. The population has more people of reproductive age than it does older people, and even more young people who haven't yet reached reproductive age. When the young people reach reproductive age, the number of people having children will increase. Thus, the wider the base of the pyramid relative to the top, the faster the population is growing.

    An age-structure diagram that is more columnar represents a population that is no longer growing. The number of people of reproductive age is about the same as the people who haven't reached sexual maturity. The number of older people, who are beyond their reproductive years, is a little bit less because people are reaching the end of their lifespan.

Examples of Different Age Structures of Populations

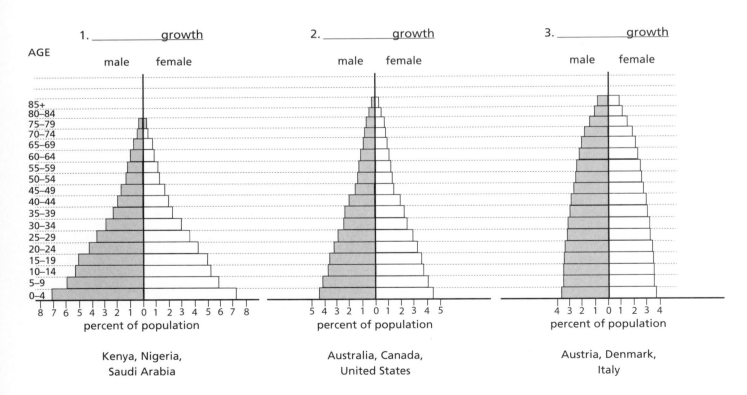

1. _____ growth — Kenya, Nigeria, Saudi Arabia

2. _____ growth — Australia, Canada, United States

3. _____ growth — Austria, Denmark, Italy

Answers

1. rapid, 2. slow, 3. zero

Threats to Biodiversity

Biodiversity refers to the number of different kinds of organisms in an ecosystem. Biodiversity is important for the health of ecosystems and also for humans. Healthy ecosystems produce oxygen, help clean air and water, have rich soils, and provide pollinators. In addition to these benefits, ecosystems with high biodiversity provide humans with a variety of food and medicinal plants, building materials, and opportunities for tourism and recreation. Also, many indigenous people have strong cultural ties to their native lands.

    Unfortunately, the growth of the human population is having many negative impacts on biodiversity. Rising global temperatures are putting stress on many species. Populations are declining due to destruction of habitats to make way for human habitation and agriculture. Human use of natural resources often outpaces reproduction rates, leading to overexploitation of resources such as trees and fish. Pollution can also damage habitats or harm species directly when they ingest human pollutants. As humans travel the globe, we may carry other species with us, introducing them to new habitats. These species may become invasive, displacing native species that are important components of food webs. Also, pathogens may become deadly problems when introduced to a new population. If these effects continue unchecked, scientists predict we will experience a very rapid loss of species in a short amount of time, leading to a mass extinction event they refer to as the sixth mass extinction.

Examples of the Threats to Biodiversity

1. _____ _____

2. _____

Asian carp

3. _____

4. _____

5. _____

6. _____

Answers

Index

Major topics are indicated with **bold** page numbers.